Iain Carson has been the Industry Editor of the *Economist* since 1994, covering the airline, transportation and manufacturing industries. He has also worked as a reporter and anchor for BBC Television and Channel Four.

Vijay V. Vaitheeswaran is an MIT-trained engineer who spent ten years covering global environmental and energy issues for the *Economist*. He is the author of *Power to the People*.

Also by Vijay V. Vaitheeswaran

*Power to the People: How the Coming Energy Revolution
Will Transform an Industry, Change Our Lives, and Maybe
Even Save the Planet*

ZOOM

The Global Race to Fuel the Car of the Future

Iain Carson
and
Vijay V. Vaitheeswaran

PENGUIN BOOKS

PENGUIN BOOKS

Published by the Penguin Group

Penguin Books Ltd, 80 Strand, London WC2R ORL, England

Penguin Group (USA) Inc., 375 Hudson Street, New York, New York 10014, USA

Penguin Group (Canada), 90 Eglinton Avenue East, Suite 700, Toronto, Ontario, Canada M4P 2Y3
(a division of Pearson Penguin Canada Inc.)

Penguin Ireland, 25 St Stephen's Green, Dublin 2, Ireland (a division of Penguin Books Ltd)

Penguin Group (Australia), 250 Camberwell Road, Camberwell, Victoria 3124, Australia
(a division of Pearson Australia Group Pty Ltd)

Penguin Books India Pvt Ltd, 11 Community Centre, Panchsheel Park, New Delhi – 110 017, India

Penguin Group (NZ), 67 Apollo Drive, Rosedale, North Shore 0632, New Zealand
(a division of Pearson New Zealand Ltd)

Penguin Books (South Africa) (Pty) Ltd, 24 Sturdee Avenue, Rosebank, Johannesburg 2196,
South Africa

Penguin Books Ltd, Registered Offices: 80 Strand, London WC2R ORL, England

www.penguin.com

First published in the United States of America by Twelve,
an imprint of Grand Central Publishing 2008
First published in Great Britain in Penguin Books 2008

1

Copyright © Iain Carson and Vijay V. Vaitheeswaran, 2008
All rights reserved

The moral right of the authors has been asserted

Printed in England by Clays Ltd, St Ives plc

ISBN: 978-0-141-03672-4

www.greenpenguin.co.uk

Penguin Books is committed to a sustainable future
for our business, our readers and our planet.
The book in your hands is made from paper
certified by the Forest Stewardship Council.

For Sarah and Michelle

Contents

ZOOM

The Great Awakening

Oil is the problem; cars are the solution

If you want to see the future of automobiles and energy, you don't need to travel to Japan. Look no further than Troy, Michigan, where a latter-day Thomas Edison is forging the path.

"The ages of mankind have been classified by the materials they use—the Bronze Age, the Iron Age, the Age of Silicon. We are at the dawn of the Hydrogen Age." So proclaims Stanford Ovshinsky, cofounder of Energy Conversion Devices (ECD). "What is more, the hydrogen economy is happening already."

It is refreshing to find a hydrogen advocate who has actually come up with the goods. After all, plenty of grandiose but unsubstantiated claims have been made over the past few years about the potential for hydrogen to replace fossil fuels as an energy carrier, so some skepticism is certainly in order. In particular, George Bush and the big car manufacturers have crowned fuel cells as the long-awaited replacement for the internal-combustion engine, but the date of commercialization for those automotive fuel cells somehow keeps slipping just beyond the horizon. Many argue that hydrogen is just a cynical long-term diversion used by Bush and Detroit to avoid short-term action on fuel-economy standards, plug-in hybrids, or other here-and-now options.

And yet, here is Stan Ovshinsky, still trumpeting the virtues of hydrogen. Ovshinsky is no hydrogen hypester. He first outlined his vision for what he calls a hydrogen loop some five decades ago as an alternative to fossil fuels. It starts with water, broken down by solar-powered electrolysis into useful hydrogen fuel that is stored in solid form or in batteries. That hydrogen is then used to power fuel cells, which release nothing but perfectly pure water vapor back into the atmosphere: "The loop goes from water to water!" he explains. Ovshinsky's green credentials are impeccable. He and his wife Iris founded ECD in 1960 with the goal of "using creative science to solve societal problems." They foresaw—long before the oil shocks of the 1970s—that the world's addiction to oil would lead to unacceptable side effects, ranging from resource wars to climate change.

Unlike hydrogen dreamers, he had actually developed products that overcome many (though not all) of the obstacles facing hydrogen. Billions of his breakthrough nickel-metal hydride batteries are now found in consumer devices, ranging from cell phones to game devices to laptops. The solid hydrogen storage system in his batteries is also used by Toyota in all of its Prius models, as well as in hybrid cars made by General Motors (GM), Honda, and other manufacturers. In other words, his quiet but dramatic work on hydrogen energy systems has made the single most important advance in automotive technology—hybrid electrics—commercially possible.

Ovshinsky's vision describes an arc that takes us from hydrogen batteries today to those superclean hydrogen-fuel cars twenty years into the future. He has delivered the key technology that is getting alternative cars on the move; his vision offers the ultimate solution, if it can clear the hurdles in its path. All along this trajectory, alternative engines and fuels will come and go as car and energy companies vie in the race to be first with the new technology that will serve the needs of a world worried about energy security and global warming. *Zoom* will tell you how the world came to be faced with this particular challenge, how we got into such a mess. It will chart the future challenge as billions of Chinese and Indians get wheels. It will mark your card on the runners and riders in the new global race to make the cars of the future.

You might wonder why this earnest and innovative do-gooder is fiddling around, trying to improve the automobile. After all, aren't cars the problem in the first place? Cars are often scorned as the most intractable source of global warming. Our utter reliance on automobiles exposes the economy to potential economic shocks from volatile oil prices. To judge by the backlash against gas-guzzlers in recent years, many Americans seem to be convinced that sports-utility vehicles (SUVs) are the work of the devil. So perhaps it makes sense to get rid of the problem altogether.

The Real Trouble with Cars

Just imagine a world without cars. Suddenly, it might seem that three great evils widely associated with automobiles—environmental harm, economic pain, and geopolitical insecurity—would vanish. But realistically, a world without cars would be a dim, joyless place with much-diminished freedom, mobility, and prosperity.

This is especially true for America, the birthplace of the modern automobile industry and most of the policy, technological, and cultural developments behind mass motorization. From drive-through banks, to drive-in churches, to roadside Holiday Inns, car culture permeates American life.

After a century of motorization, America crossed a threshold in 1995, when cars and light trucks first outnumbered driver's licenses in the country. Today, there are often more cars in the average American household's driveway than there are drivers inside, and three-car garages are becoming the norm. Inspired by the American example, such developing giants as China and India are now taking to the roads too. Soon, we will be a world of a billion cars.

Oil is the problem, not cars. That is why we must reinvent the automobile. As engines of change, the clean cars of the future can help speed the world toward a more sensible approach to transportation. The snag is getting from here to there. Big Oil clearly has no interest in seeing its main product fall by the wayside, and the Detroit car industry has shown few signs of real innovation or long-term vision.

Stan Ovshinsky and a growing band of entrepreneurs, innovators, and outsiders are now working furiously to spur the sorts of innovations that the established industry powers, Big Autos and Big Oil, simply refuse to develop. The incumbents are doing precious little to save the planet, only perhaps changing a little to kill it less quickly. The energy industry has long had the knowledge needed to pursue petroleum substitutes but has chosen to defend and milk its existing gasoline assets. The car industry has also had the technical ability to produce cars with much greater fuel efficiency but has chosen to build gas-guzzlers instead. Believe it or not, Henry Ford's Model T was a "flex-fuel" car that could run on either ethanol or gasoline and got better gas mileage than the average new vehicle sold in America today. Worse yet, both industries have bitterly fought government efforts to encourage the development of more efficient cars and alternative fuels or co-opted and corrupted such efforts to the point that they became meaningless.

Now the modern world's most important energy visionary believes that we are on the cusp of a clean-energy revolution. Ovshinsky's vision for a hydrogen loop was just a blackboard exercise when unveiled five decades ago. But since then, he has invented a new field of science (amorphous materials, named Ovonics in his honor) and produced innovations to bring that loop closer to reality. Joachim Doehler, a senior scientist at ECD, says, "Stan starts with a vision—say, 'The computer must work like the mind does'—and then goes out to invent what we need to get from here to there." It is a systems approach.

The best evidence of this at work is Ovshinsky's new solar factory in Michigan. Several decades ago, he argued that solar panels must be made not as brittle crystalline panels in costly batch processes—how almost everyone else does it today—but "by the mile." He was ridiculed. Ovshinsky refused to yield, demanding his team come up with processes for producing miles of thin-film solar material. Doehler, a veteran of AT&T's legendary Bell Labs, recalls telling him it was impossible. The boss proved him wrong, personally designing much of the solar factory from scratch using unusual industrial gases.

Some challenge his record. An article in *Forbes* asked in 2003 why investors "keep giving money to an inventor who can create anything but profits." ECD has lost money for most of the forty-plus years it has been public. As even one of his loyal lieutenants confesses, "This company would have gone bust six times already if it were not for the personal loyalty people felt for Stan; we went the extra mile because this place is unique."

That revelation points to one big way in which the two great geniuses of energy differ. Thomas Edison was a hard-charging man whose pursuit of profit, product, and prosperity drove GE to commercial success. Contrast that with the Ovshinsky vision and ECD's corporate culture. "We're here to change the world. No more war over oil," he has argued for decades. ECD is clearly committed to clean energy—and Ovshinsky is clearly not motivated by money. The *New York Times* analyzed executive pay in America and found that heads of companies typically get five hundred times the salary of the average worker at their firms; the ratio at ECD is five to one.

The Next American Revolution

A powerful grassroots movement—call it the Great Awakening—is clearly under way that is sparking a great global race to fuel the car of the future. People are increasingly fed up with the car companies and oil titans, and they are all too aware that these industries have lobbied politicians into gridlock over energy policy. As Ovshinsky and his upstart collaborators were busy developing innovative hydrogen technologies for cars, a separate nationwide coalition of companies and local governments was coalescing around the idea of "plug-in" hybrid cars. Unlike the hugely successful Toyota Prius, a hybrid-electric car that can never be plugged in, these hybrids charge up a battery overnight that can power the first 20 or 30 miles of travel each day. The rest of the day's travel would use gasoline (or, in the future, hydrogen or ethanol fuel). Fed up with the federal government's refusal to enact mandatory curbs on greenhouse gases, local and state governments are forging ahead with their own. Revealingly, today's push for cleaning up carbon and cars is a bipartisan movement: its leaders are Republicans like California's Arnold Schwarzenegger and New York's George Pataki, as well as Democrats like Iowa's Tom Vilsack and Illinois's Barack Obama.

The world is at an energy crossroads, and the decisions made about cars and oil in America and China over the next decade or so will set the course for the coming century. That is because energy infrastructure can last for decades, and the carbon emitted can last even longer. If we are to set our energy system on the right course before real crisis hits in a decade or two, we need to start that transition now.

Even Detroit could yet prove to be the automotive superpower of the twenty-first century, if Ovshinsky has his way. He is an inspiration in his own right, but there is a larger lesson to be drawn from his extraordinary life that is relevant to the future of cars and oil. And despite the persuasive case he makes for it, that lesson is not that the future belongs to hydrogen and only hydrogen. That may well turn out to be right, especially when the hydrogen economy is defined not merely as automotive fuel cells but as a holistic "water to water" energy loop of the future. The history of energy over the centuries has been decarbonization: man progressed from burning wood and peat to using coal, oil, and natural gas and now to hydrogen. In doing so, we are shifting from carbon-heavy, dirty hydrocarbons to hydrogen-heavy, cleaner ones.

Take the very long view, and hydrogen is clearly the ultimate zero-emissions energy carrier. Even so, the more important (and certainly more immediate) lesson to be drawn here is this: Detroit may be down, but it does not have to be out. And it does not have to bow to Japanese or other foreign rivals. For proof, look no further than Akron, Ohio.

Most Americans probably consider Akron the most ordinary of cities, but not Don Plusquellic. The city's long-serving mayor has a vision for transforming Akron from a washed-up industrial wasteland into a city of the future. A few years ago he thought of putting solar panels on the roof of the National Inventors Hall of Fame, housed in the city. As he looked into that project, he came across the Ovshinskys. "Meeting them had a huge impact on me," he told Ovshinsky's biographer, George Howard, a professor at the University of Notre Dame. "Seeing Stan's hydrogen loop made me think I had actually seen the future. I thought, 'Why not Akron?' Let's get into the hydrogen business."

Why not indeed. It is not only far-flung island states—ranging from Iceland to Hawaii to Vanuatu—concerned about pricey petroleum imports that can embrace a clean-hydrogen future. And it is not just places with hypergreen economies like California and Sweden that can declare a firm goal of becoming completely independent of petroleum. America's industrial heartland can leapfrog ahead too. ECD and Akron have started with hydrogen testing and storage, but the plans go much further. Using the ample water that the city once used to make rubber (back when it was the tire capital of the world), Ovshinsky plans to make hydrogen using the abundant, cheap, off-peak electricity available from the grid utilities at night. He saw potential for using the city dumps as a source of both biofuel to produce electricity and methane that can make hydrogen: "We can make Akron the 'Saudi Arabia' equivalent of the world . . . it'll be a working example of what needs to be, and can be, accomplished throughout the United States."

The key obstacle now is Washington's backward-looking, obstructionist approach to energy—a pork-barrel fiesta that Senator John McCain has called the "leave no lobbyist behind" approach. That has led some to despair that nothing good can ever come out of Congress on energy, given the power of the oil and car lobbies. Techno-utopians argue that magical new technologies will save us, while market fundamentalists say that the invisible hand will do the trick. Well-intentioned corporations keen on clean energy and carbon-free technologies make the argument that "corporate social responsibility," not public policy, is the key. And small-government types are anyway suspicious of Washington.

Even the eternally optimistic Ovshinsky seems to share such doubts. "Forget great ideologies! Politics can't live up to its promise of a better world—but value-driven science and technology can, by improving the material base. The key is practically improving the lives of ordinary people."

Here is why all of these groups are wrong. When it comes to the thorny geopolitical, environmental, and economic complications involved with cars and oil, America's federal energy policies do matter. The heady mix of perverse subsidies for fossil fuels and the absence of proper "externalities" taxation of gasoline leaves the game rigged in favor of Detroit and guarantees continued oil addiction. This will not change magically unless the incentives facing entrepreneurs and innovators change: clean technologies will not get their just rewards in the marketplace, and new markets for carbon-free energy will not take off, unless we fix what's wrong with energy policy so that the playing field is level.

After all, the business of business is business—as it should be. Contrary to what some critics claim, there is nothing inherently evil about oil companies pumping oil or carmakers selling cars. That is, in fact, their job—and for decades, it was socially acceptable for them to do so. The difference today is that society's expectations are changing: a richer, greener, better-informed world is demanding much more from its energy and transportation industries. The social contract is evolving—but public policies have not yet changed to reflect that progress.

That, in sum, is why government still matters. Only sensible and courageous action by government to take account of the external costs of burning oil can set things on the right course. Those external costs are not reflected in the pump price of gasoline, but of course we pay for them through the Pentagon budget, the suffering of asthmatic children, and the pain of economic shocks. Only if the federal government spurs change, either through market-based regulation or better yet through carbon taxes, will we level the playing field and give clean cars and carbon-free energy a fighting chance.

That will happen only if everyone is ready to abandon the myth of cheap fossil fuels and pay an honest price for gasoline. There are signs that the Great Awakening is changing consumer attitudes on this crucial issue too. Ask ordinary Americans if they will simply support a hike in gasoline taxes, as the *New York Times* did in 2006, and the majority say no. But when the pollsters asked whether those same people would be willing to support higher gasoline taxes if the money went to reduce oil imports or to fight global warming, a strong majority responded yes.

That suggests that the country is just ripe for a new approach to this issue. Americans will follow political leaders with vision and courage, who put forth a comprehensive, bipartisan, long-term strategy to tackle oil addiction and global warming. No one likes to pay taxes, but Americans do respect straight talk and have always had a strong sense of fair play. If political leaders take the trouble to explain the notion of oil's external harms and make the case for subsidy and tax reform, citizens will rally to the cause.

And if the next president and Congress really do embrace an innovative policy and stop propping up the tired old giants of the fossil-fuel and automobile businesses, then radical change will be possible. Entrepreneurs and innovators would then ramp up their investments, and we could see a technological revolution that makes clean, efficient, gasoline-free cars possible so that the developing world's legitimate needs for energy and aspirations for mobility can be satisfied, while the rich world's concerns about the environment are met too.

But Washington, D.C., will act only if ordinary Americans—voters, consumers, drivers one and all—speak up, step out, and demand an end to business as usual. As Saint Thomas More argued five centuries ago, government is simply too important an enterprise to leave to the scoundrels; it is the duty of honest, everyday folk to get involved and to make sure our country heads in the right direction.

By taking the real problems posed by cars and oil seriously while debunking wild-eyed claims made by the chorus of despair, the authors hope this book will serve as a call to arms. The challenges are daunting, but the solutions are within grasp if readers mobilize and energize the political process in favor of clean energy. Indeed, there is every chance that they can turn this crisis into opportunity, transforming the grease and grime, soot and sulfur industries that built the twentieth century into the clean, sustainable building blocks of the twenty-first century.

The global race to fuel the car of the future is on.

I

HIGHWAY ROBBERY

America's car and oil industries got the world into the current energy crunch and in the process may have killed off their own future

CHAPTER ONE

The Terrible Twins

Cars and oil wrote the history of twentieth-century American capitalism

After a century of prosperity and power, the industries that shaped America more than any others are now at a crossroads. The age of oil and cars is giving way to something new. Together these two industries dominate world business because of their sheer size. The big five of oil—ExxonMobil, Royal Dutch/Shell, BP, Chevron, and ConocoPhillips—and the world's automobile giants—General Motors, Toyota, Ford, Daimler, and Volkswagen Group—dominate the lists of the global top-fifty corporations. Each has sales between $100 billion and $200 billion.

They spend huge sums each year looking for oil, extracting it, making today's cars, and developing technologies for the cars of tomorrow. GM in the past five years has spent the seemingly awesome sum of $1 billion on research into clean cars with hydrogen-fuel-cell engines, with no prospect of making a profit on this in the foreseeable future. But even $1 billion is barely two days' revenue for a company GM's size. That shows how high the stakes are now, as the car industry faces a revolution.

The really big spenders in global research and development are not the white-coated scientists of Big Pharma and biotech or the geeks of Campus Microsoft. They are the legions of engineers, chemists, physicists, and geologists employed by the oil and auto industries. The automobile manufacturers are the biggest spenders worldwide on R & D. As pressure now mounts for them to make cleaner cars, the pace of the race for new technology is picking up.

Since even America has woken up to the threats of climate change caused by the rising emissions of global-warming gases, the car industry is in the front line of the battle against carbon. About a quarter of the man-made greenhouse-gas problem comes from surface transportation, including ships and trains. (Air travel adds another 3 percent.) Cars and light trucks make up the lion's share of such mobile pollution sources. As the oil industry starts tapping the vast but carbon-intensive reserves of "unconventional" hydrocarbons like Canada's tar sands and America's shale, which can be converted into gasoline at great environmental cost, the impact of transportation on global warming may rival even that of coal-fired electricity production.

What is more, because everyday life in most rich countries is built largely around the car, it is sure to be the most difficult source of carbon emissions to moderate. Over its decades of fighting smog, Los Angeles has discovered that it is really hard to change individual behavior to reduce consumption of gasoline, and so it turned to techno-fixes like catalytic converters, which reduced the emissions associated with such gas-guzzling instead. That painful experience explains why technological solutions to tackle the carbon from automobile tailpipes are the key.

Today the problem is greatest in America, Europe, and East

Asia; these three markets account for three-quarters of the greenhouse gases pushed out the tailpipes of the entire global new-car fleet of more than sixty million cars and trucks each year. On average, American vehicles emit around 480 grams per mile, just over double the European level. For Asia, the figure is halfway between the two. The European figure has come down because of the increase in the use of inherently more fuel economical diesel engines, which now power about half of all new cars. The rise of diesel, in turn, is itself the result of European governments placing heavy taxes on gasoline (the retail price for a gallon at the European pump can be two or three times that paid in America) and lower taxes on diesel.

That European models such as Mercedes are now being launched in America with diesel engines is one early indication of technological improvement spreading around the world in response to the global-warming challenge. But that is incremental and inadequate; the scale of the global oil and carbon challenge demands a much bigger response. Any such revolution must be on a huge scale because of the sheer size of the global auto industry.

And Amory Lovins, a farsighted energy thinker who lives atop a mountain in the Rockies, has been arguing for some years now that the car and energy industries must embrace radical change—and, much to the annoyance of industry bosses, has been showing them how to do it.

Ripe for Revolution

It is a rare company prospectus that begins with a quotation from Goethe: "Whatever you can do, or dream you can, begin it. Boldness has genius, power, and magic in it." But Lovins is not a normal entrepreneur, as anyone who has met this eccentric and disheveled but unmistakably visionary thinker knows.

The founder of the Rocky Mountain Institute, a leading green think tank based in Old Snowmass, Colorado, thinks the car industry's incremental approach to cutting emissions and improving fuel efficiency will never amount to much. He wants a complete redesign of the automobile, from the bottom up, and intends to show the big boys how it should be done.

This is not the first time the Sage of Snowmass has challenged conventional wisdom. Back during the gloom and doom of the 1970s oil shocks, most energy pundits were convinced that energy consumption and economic growth could proceed only in lockstep, thus making scarcity and future shocks inevitable. In a controversial article in *Foreign Affairs*, Lovins argued back then that there was an alternative "soft" path: if governments, companies, and individuals embraced energy efficiency and other demand-related approaches, then economic growth could be decoupled from gas-guzzling. He was widely ridiculed at the time as naïve or worse, but history has vindicated him. Thanks to public policies like gasoline taxes in Europe and automobile fuel-efficiency standards in America, the world did indeed embrace a soft energy path, and another oil shock was avoided for two decades.

But as the memories of those earlier oil shocks have faded in America, so too have those virtuous policies. The American economy is now much less energy efficient than its chief international rivals, and the average fuel economy of new cars in the country is close to a twenty-year low. Even Henry Ford's Model T got better gas mileage a century ago than today's average new car! The oil and car industries may be spending a fortune on R & D, but their mind-set remains incremental and risk-averse. They are clearly not innovating with the vision and verve they showed back during the golden age of the automobile a century ago.

But now, Lovins has devised a concept car that he hopes will spark a revolution in the motor industry and "revive the spirit of Henry Ford and Ferdinand Porsche." For a decade, a crack team of engineers brought together by Lovins beavered away in a hideout high in the Rocky Mountains. They came up with the Hypercar, a sleek new automobile powered by a fuel-cell, zero-emissions engine. This engine takes oxygen from the air and hydrogen from its tank to create a chemical reaction that produces electricity and water, the only by-product.

This alone would be unremarkable, given that all the world's carmakers are now into fuel cells. The difference with the Hypercar vision is that it takes a holistic approach. The entire body is to be made of composite plastics. The transmission and steering are entirely electronic, which removes the need for clunky mechanical parts. Instead of a steering column and wheel, there will be game-machine joysticks, as in the cockpit of the latest Airbus jets—where a "fly-by-wire" system largely replaces heavy hydraulic and mechanical controls. The result is a big car with a fuel economy comparable to 100 miles per gallon of gasoline, far higher than today's most economical diesel or hybrid-electric cars can achieve.

Can this be serious? Actually, the technologies that Lovins champions are not really far-fetched. Carbon composites, electronic controls, and even fuel cells are feasible today. The reason they have not been much used in cars is that established carmakers have invested vast sums in conventional manufacturing technology, plants for stamping old-fashioned steel assemblies, and the like. This has made them reluctant to embrace radical approaches; they pooh-pooh the Hypercar as fanciful and irrelevant. Undaunted, Lovins has put the key concepts for his car up for grabs as open-source material, hoping that fresh thinkers outside the conventional car industry will pick them up. He has also set up a new company, Fiberforge, that is promoting advanced composites to the critical suppliers of parts and assemblies to the automobile industry; if they adopt his ideas, Lovins may be able to do an end run around the incumbent giants.

What makes him so tenacious? One reason is that Lovins knows from experience what it means to take on and defeat a giant, well-connected, and obstinate global industry. He cut his teeth as an environmental thinker during the fiery antinuclear battles a few decades ago, working closely with David Brower, the founder of Friends of the Earth (immortalized by John McPhee in *Encounters with the Archdruid*). But a deeper motivation is his desire to change the world for the better, which he inherited from his parents. "They taught me always to strive for self-improvement and to help others," he says with characteristic modesty. Despite his recognized genius, Lovins frequently gives credit to past mentors, praises rivals and predecessors alike, and often quotes great thinkers. "Sometimes having a new idea simply means stopping having an old idea," which he attributes to Edwin Land, is one of his favorites.

With that philosophical bent and a mischievous tendency to poke at authority, Lovins was destined to be a troublemaker in whatever field he ended up going into. That should have been physics, given his natural abilities in that field, but energy captivated his attention. He became convinced early on that global warming was going to be a major problem (he wrote his first paper on it back in 1968) and saw energy as "the master key for leveraging changes in the world." When Oxford University refused to let him write his doctoral thesis on energy issues, he dropped out to pursue his investigations on his own. That spirit of rigorous independent inquiry and an endearing earnestness are still evident in the graying Lovins: "Why do I keep doing this? Because I've got all this curiosity and an atticful of knowledge that I'd like to apply to improve the world."

Dreamer though he may be, even Lovins is enough of a realist to accept that radical change will not come easily, given the sheer complexity of the product involved. The everyday automobile is a blend of art and science so elaborate, he acknowledges, that "it is beyond baroque: it's rococo."

Beyond Baroque

The auto industry is so big and complex that it is different from any other manufacturing enterprise. The industry's 60 million vehicles produced per year consume the lion's share of the 85 million barrels of oil produced every day around the world. It employs millions around the world and accounts for $1 in $10 of the American economy. The management thinker Peter Drucker dubbed autos "the industry of industries." For a hundred years, it has been more than that: the automobile is capitalism on wheels.

It may not seem like it now, with Detroit mired in losses and losing ground to Japanese, South Korean, and European competitors every year. The stars of capitalism today work in the technology and media industries. But a hundred years ago, the exciting young industry was automobiles, and the emerging stars of free enterprise were no longer the nineteenth-century barons called Rockefeller, Carnegie, Mellon, or Morgan.

They were the band that followed in the footsteps of Henry Ford. There was Willie Durant, Wilfred Leland, Ransom Olds, the Dodge brothers, Charlie Nash, and Louis Chevrolet. They all started little car companies that would become part of the Microsoft of its day: General Motors. Durant started his business life as part owner of a roller-skating rink in Flint, Michigan. He moved into the nascent auto industry, forming

GM in 1908, and proceeded to buy up twenty-five small start-up car companies in the space of eighteen months. He lost control to his creditor banks after two years, only to win it back a few years later, before being squeezed out in 1920, when the legendary Alfred Sloan took the wheel. Durant lost a fortune in the Wall Street crash, declared bankruptcy in 1936, and ended his days running a bowling alley back in Flint, with scarcely a dime to his name. Along the way, he had founded the company that was to lead the industry. By 1926, Ford was overtaken by GM, as Sloan realized that customers wanted a variety of models, not plain-old, utilitarian Tin Lizzies.

Another star of the young auto industry was Walter P. Chrysler, a former railroad engineer who by 1916 became the best-paid executive in the auto industry, working at Buick and later Willys. He was a larger-than-life character. He bought a huge estate at Great Neck on Long Island, with a 30-foot swimming pool, an eight-car garage, and a waterfront of 450 feet on Long Island Sound, equipped with a 150-foot pier to reach his yachts. By 1925, he had his own car company, the Chrysler Corporation, born of the ailing Maxwell Company, which he had revived using his manufacturing skills to develop the 1924 Chrysler car. "I gave the public not only quality, but beauty, speed, comfort in riding, style, power, quick acceleration, easy steering, all at a low price," he said.

Heading for Divorce

Today the car industry in America seems at first sight a million miles from such glamour. It looks more like a pension system with a car and credit business on the side. Its relations with its lifelong partner, Big Oil, are also looking tattered, as the auto industry seeks alternative energy sources.

This is leading to an outpouring of innovation and entrepreneurship. High-tech millionaires in Silicon Valley think Big Auto and Big Oil could be unseated by disruptive technology. Tesla Motors, a California upstart funded by the men behind Google and PayPal, has picked up on Amory Lovins's ideas on lightweight composites and come up with a sizzling, all-electric sports car that is greener than a Prius and faster than a Ferrari.

That is impressive, but there's more to the coming revolution than a few rich dreamers on the West Coast. Consider the sorts of things popping up all round the world. The most visible in America, Japan, and Europe is that Toyota gasoline-electric hybrid, the Prius, which transforms fuel economy by hitching battery electricity to the internal-combustion engine—the best example of the traditional industry embracing new thinking. With a hybrid system, a small gas engine packs the punch of a big one without paying the price of lower gas mileage. But enthusiasts in California have already discovered how to hack into these cars so they can also be plugged into the grid overnight. Everyday motoring of up to 30 miles can be done purely on electric power, with the engine only a reserve in case the battery runs low. These hackers are even forcing Toyota to move ahead with better batteries so that standard plug-in hybrids can roll off its assembly lines.

Buses have been spotted on the streets of Seattle, Chicago, and other American cities that produce nothing out of their tailpipes but clean water vapor. They use hydrogen fuel cells, which make electricity by mixing the gas with oxygen. GM is working on a mass-production, fuel-cell sedan that it could bring to market in 2010, if enough gas stations get around to supplying hydrogen, extracted on-site from piped-in natural gas. The streets of Shanghai have become so clogged with cars and the air so thick with pollution that locals are turning to smart electric bikes. On the outskirts of the city, China's own Detroit is not just making conventional autos but is working with hundreds of labs around China to leapfrog from polluting internal-combustion engines to fuel-cell cars that emit no pollution or carbon-dioxide greenhouse gas to add to global warming. In Brazil, virtually all new cars can run not just on gasoline but also on ethanol made cheaply in that tropical country out of sugarcane. The midwestern states are pushing hard for such biofuels to be made available across America, with the ethanol made out of corn. One day, plant waste might become an economical source of ethanol, one that won't need another raft of farm subsidies.

Toyota is beating Detroit by being the first big automaker to make great strides in fuel economy. European firms such as Mercedes and BMW are pursuing a parallel path with economical new diesels that don't pollute like the dirty old puffers Detroit used to make: because they get many more miles per gallon, they contribute less than gasoline cars to global warming. In India, one of the world's newest volume carmakers, Tata Motors, part of a huge family conglomerate, is now working furiously on a revolutionary $2,000 car with fantastic gas mileage, perhaps three times that of the average new American car.

There has not been such effervescence in autos since the early days of the industry, when nine out of ten autos were battery powered. Over ten years, internal-combustion vehicles gradually began to assert themselves over steam and electric cars, winning a third of the market. Ten years later, helped by the invention of the electric-starter motor, the internal-combustion engine won the race because gasoline packed more power per gallon, though Henry Ford was canny enough to ensure his Model Ts could also run on the ethanol that midwestern farmers could brew from corn. Now a similar power struggle is taking place to power the car of the future.

As the green revolution gathers pace, it will force change on the oil and auto industries, rendering useless hundreds of billions of dollars' worth of assets. No wonder both industries have fought to block progress on environmental advances. No wonder Texas and Detroit are scared that the initiative has been seized by nimbler foreign rivals, from the automakers of Asia to the energy giants of Europe. The meltdown that Detroit has been going through for a decade is changing America. Henry Ford not only gave Americans wheels, he gave the world its first (and possibly last) blue-collar middle class. In return for backbreaking, monotonous work on his assembly lines, he paid wages good enough to allow working men to own homes in the suburbs that were springing up because of the transportation freedom brought by the car. Consolidation of the industry into the Big Three preserved this good life until free-trade globalization brought foreign competition.

Robbed of their oligopoly power, GM, Ford, and Chrysler got into financial difficulties that still dog them. Tens of thousands of workers in the midwestern states of Michigan, Illinois, and Ohio lost their jobs and were left with little prospect of finding equally well-paying alternatives. This alone is one of the reasons much of Middle America is standing still economically. It goes a long way toward explaining how lower-middle-class incomes have been left behind in the explosion of wealth that new industries and businesses have brought to millions of Americans.

Automobiles, the Love Affair of the Century

Modern America and the automobile grew up together. But like so many early-twentieth-century Americans, it was born elsewhere. "Automobile" is a French word. The nursery of the industry was France. In 1900, just seven years after the French people gave the United States the Statue of Liberty as a belated centennial birthday present, René Panhard and his partner Emile Levassor gave us the modern motor car. They defined the way it looked before Henry Ford. They arranged the *système panhard*, the basic fore-to-aft formation of radiator, engine, clutch, gearbox, prop shaft, and rear axle. This was far removed from the German Daimler-Benz horseless carriage, which had a gasoline engine turning a belt drive to the wheels.

Serendipitously, the French flavor went further: half of GM's brand names, so etched into the minds of Americans, are French. Detroit (it means "narrows" in French) was founded by Antoine de la Mothe Cadillac, a French soldier fighting the British colonists who later governed Louisiana. Cadillac died, disgraced, after a spell in the Bastille prison, little knowing how his name would live on. Louis Chevrolet was a Swiss-French racing car driver who came to the United States to drive a Fiat in an early auto race in New York in 1905. He formed the Chevrolet Motor Company and never went back. Then there's Pontiac—the French name of the chief of the Ottawa tribe of Native Americans. All these French names became synonymous with American capitalism as the auto industry accelerated the twentieth century into unheard-of prosperity.

The 1920s marked the emergence of the modern auto industry. Will Durant put together Olds, Buick, Chevrolet, and Cadillac to form GM, which in 1926 outsold Ford. When he was struggling with the banks, it was another bunch of Frenchmen, the DuPont clan (immigrant French aristocrats who had become the leading industrial family of late-nineteenth-century America, making gunpowder), who bailed him out, making a second fortune from their investment. Henry Ford gave the world the technology for the cheap, rugged car—you could have any color you wanted, he loved to quip, so long as it was black.

Durant's GM, especially after he was replaced at the top by Alfred Sloan in 1920, took the opposite tack. Sloan's idea was to offer a selection of models, one for every purse. Styling would differentiate one year's cars from the next year's crop. So buyers would be encouraged to trade in their old car for a new one that demonstrated their growing affluence. Both Ford and GM opened credit businesses so that ordinary Americans no longer had to save up to buy their autos: installment purchasing, invented in 1915, was at the heart of the burgeoning industry. Assembly-line production raised efficiency and wages—up 26 percent in real terms between 1920 and 1929.

Others followed Henry Ford in paying good wages for the hard, repetitive work. By the end of the decade, output from Detroit had gone from 1.9 million cars in 1920 to 4.5 million in 1929. Automaking had become the biggest industry in the country. It launched another industry in the form of the dealer networks that spread across all forty-eight states to sell cars. Car dealers became big players in each local business community and gained great political clout at the state level, as witnessed by the laws that preserve their franchises.

The Wall Street crash and the Great Depression sent car production down to a quarter of its boom production. It would not regain its 1929 peak until after the austere years of World War II, in 1949. During the war, Detroit became Bomber City, as its factories turned to meeting military needs for warplanes and three million trucks and tanks for the war in Europe and Asia. It was the first motorized war. After the Normandy landings by the Allies, 3-ton trucks designed by GM in Detroit were carrying German soldiers (in Opel-badged vehicles), American

GIs in Chevrolets, and British troops in Bedfords. GM's German subsidiary, Opel, even shipped dividend checks to its parent throughout the war.

With the war ended, ten million veterans headed back home, and the auto industry began its remarkable boom. In the decade from 1945, the number of cars on American roads doubled to fifty million, and Detroit was working flat out, producing three million cars a year. Returning GIs wanted wheels and an individual home out in the suburbs, away from the grime of crowded cities. Ironically, the same thing is happening in China right now, as the ordinary Chinese gain the freedom to buy homes and as crowded downtown apartment buildings are razed for new office developments in the booming cities of the coastal eastern provinces.

The Asphalt Age

Just as the auto was the symbol of the Roaring Twenties, so it became the icon of the new era of postwar affluence, that long boom that went from the peace of 1945 until the Yom Kippur War of 1973 and the first oil shock. While Europe shivered in the ruins left by the war, and while China followed thirty years of war with a bloody revolution that brought the communists to power, America saw an age of affluence that reached much more of the population than did the fitful flare of the 1920s. Real incomes went up by nearly a fifth between 1947 and the start of the 1960s. Income-tax relief on 100 percent mortgages put the single-family suburban house in reach of millions. A deposit of $100 and three years to pay put a cheap car in the driveway. By the late 1970s, there was one car for every two Americans. Three decades later, there are over 250 million cars for 300 million Americans, and the three-car garage is standard in many suburbs.

All these cars needed roads. Federal funds were freely available to top up state spending on new roads to carry the growing tide of traffic. Starting with Detroit, New York, and Chicago, new, fast routes were constructed to clear access to the city from the suburbs. By the early 1950s, cities such as Philadelphia, Pittsburgh, Baltimore, New Orleans, and Boston did the same, as 12,000 miles of such urban expressways were built. Meanwhile, rail-track mileage shrank to half what it had been at its peak, in 1916. The apogee of the Asphalt Age was the creation of the interstate highway network, crisscrossing America from coast to coast, from Canada to Mexico. In June 1956, President Dwight Eisenhower signed the act that created the National System of Interstate and Defense Highways at a total cost of $100 billion, financed out of gas taxes.

It still stands as the biggest peacetime government investment program. The Pentagon and defense industry had long lobbied for roads to link military bases in inaccessible places. Propaganda posters of the time emphasized how useful the roads would be for mass evacuations in the event of nuclear strikes by the Soviet Union. City leaders signed up for the program when they learned that as much as half of the cash would flow through the coffers of city halls. No wonder Charles E. Wilson (nicknamed Engine Charlie), president of GM, proclaimed in 1957, "What's good for the country is good for General Motors, and vice versa." The automobile had become the essential tool for modern living.

With the thousands of acres of new suburbs came suburban shopping malls, supermarkets, and freeways. Between 1950 and 1970, the number of suburban shopping centers grew from a few hundred to over three thousand. Holiday Inns started springing up everywhere by the side of the new roads. At the huge, sprawling junctions of the interstates, up sprang truck stops, motels, warehouses, gas stations, and diners. Soon it was possible to cross America without seeing one single city center.

All this happened just as the mechanization of farming was causing a decline in employment all across the southern states. Soon, thousands of poor black families were cramming into old jalopies or hauling battered suitcases onto Greyhound buses to take the interstate highways up north to find jobs in cities such as Chicago, Detroit, Cleveland, and New York. Seeking cheap accommodations, they moved into areas blighted by the carving out of great expressways.

It is tempting for many environmentalists, sentimentalists, and liberals to lay blame for the negative effects of this huge transformation of the face of America on the hood of the automobile. But the fact is, as even Jane Jacobs (a widely respected analyst of American cities) argued, autos per se were not to blame for the decline of cities. She wrote (in *The Death and Life of Great American Cities,* 1961), "The present relationship between cities and automobiles represents one of those jokes that history sometimes plays on progress. The interval of the automobile's development as everyday transport has corresponded precisely with the interval during which the ideal of the suburbanized anti-city was developed architecturally, sociologically, legislatively, and financially." She maintained that automobiles themselves are "hardly destroyers of cities."

Late-nineteenth-century city centers were congested, filthy, smelly, and dangerous dens of disease because of the carpet of horse manure everywhere underfoot. The only mistake, wrote Jacobs, was to replace each horse with half a dozen mechanized vehicles, instead of using a mechanized vehicle to replace half a dozen horses. Trucks and buses, in other words, would have been better forms of automobiles for cities than cars and SUVs.

The Green Backlash

If the first half of the last century was to see the auto industry, with the oil companies in tow, working to build modern America, the second half was to mark a dramatic change. Three people did more than anyone else to change the atmosphere in which the auto and oil industries worked. They were Rachel Carson, Ralph Nader, and Sheikh Ahmed Zaki Yamani. Carson's book *Silent Spring* exposed the damaging effects on wildlife of the insecticide DDT, hitherto seen as a wonder chemical, which was developed in World War II to kill off disease-spreading parasites. Its publication in 1962 gave voice to a generation of youth skeptical of big business. It was the birth of environmentalism.

Ralph Nader, as a radical young lawyer, campaigned against unsafe cars; he targeted the Chevrolet Corvair in his book *Unsafe at Any Speed*. Although the car was ultimately found, after long court battles, to be no more dangerous than any other, GM's furious and illegal attempts to harass and discredit Nader left it with the image of a corporate bully that cared more about its share price than the safety of its customers. Consumer activism was born, and the road to the seat belt, the air bag, and the crumple zone was opened.

The third agent of change was Sheikh Yamani, the crafty oil minister of Saudi Arabia during the 1970s and 1980s, who forged the oil weapon to try to control America's Middle East policy. Yamani was flying to Vienna to a meeting of OPEC (Organization of Petroleum Exporting Countries) oil ministers the very day the Yom Kippur War broke out, in October 1973. He quickly marshaled the organization, bringing a degree of discipline to its workings. The result was the quadrupling of oil prices over the next year. The emergence of consumerism, environmentalism, and an oil crisis all came together to change the relationship between America and autos. And life for the Big Three in Detroit would never be the same again.

That first oil shock, in 1973, created a panic about energy supplies and the high fuel consumption of the gas-guzzling land yachts that poured off Detroit's lines. The nascent consumer and green movements spurred by Carson and Nader had autos in their sights. Soaring car use had ruined the air quality not just in coastal Los Angeles but in most American cities, because such tailpipe gases as sulfur dioxide and carbon monoxide combined with sunlight to produce foul smog.

As pressure mounted on the auto industry, carmakers and oil companies began lobbying to oppose new federal regulations on fuel consumption, known as the Corporate Average Fuel Economy (CAFE) laws, and on rules to cut down the emission of pollutants. This effort is ongoing.

The Rope-a-Dope

Some of the lobbying is a bare-knuckles defense of short-term business interests—not much different from what the tobacco industry did for years with its influence in Washington. Two things make it more successful than the tobacco lobby, however. The first is the clever one-two dance the industries do in passing the buck; the second is a disingenuous but wily appeal to national interest.

Given that oil and cars have been joined at the hip for a century, you might think the two industries are allies. In fact, the opposite has been true since the heat was turned on in the mid-1970s. Far from presenting a united front when lobbying, the two heavyweight industries are even sometimes at loggerheads, each one pointing the finger at the other.

For example, when California's environmental regulators tried to clean up smog by cracking down on the industries, the car guys claimed the solution lay in cleaner, low-sulfur gasoline, while the oil guys said the only way forward was cleaner engines and catalytic converters. After dragging their feet for as long as possible, and after many lives were needlessly lost in the interim to pollution, the two industries were forced to take joint action.

The same "he said, she said" routine is evident in their response to global warming. The car industry hates any pressure to make cars more efficient and has fought increases in the CAFE fuel-economy laws. Oilmen hate government regulations, too, but, when asked about efficiency, Exxon's former boss, Lee Raymond, could not control his enthusiasm: "We absolutely waste too much energy; efficiency is a must!" Though he thought the idea of hydrogen energy was "stupid" and had no future, even he felt it necessary to declare that fuel cells just might replace the internal-combustion engine someday "simply because they are much more efficient." However, when it came to implementing fuel cells, his company took the bizarre (but self-serving) view that this zero-local-pollution engine should be fed dirty gasoline instead of pure hydrogen. Exxon therefore spent a huge amount of money trying to steer the country's fuel-cell efforts toward this filthiest of approaches and ultimately gave up only when its technology failed—but nevertheless, it did succeed in slowing down the hydrogen bandwagon by a few years.

Just as the car industry hates CAFE, the oil industry hates any hint of gasoline (or carbon) taxes. The industry's lobbyists have skewered any politicians who have dared to propose such taxes, which would have the inevitable effect of reducing the consumption of gasoline. But ask Ford or GM what the world should do about global warming, and they cheerfully point to carbon taxes as the smart solution. One old industry

hand even jokes, "Of course the car industry is in favor of a
gasoline tax—on the condition that it is never implemented!"
Bill Ford even made a dramatic speech after Hurricane Katrina
struck declaring a national energy crisis and called for a grand
national summit of oil and car bosses in the White House to
discuss the possibility of a carbon tax. Unsurprisingly, he didn't
get his tax.

What's Good for GM May Be Bad for America

If this one-two dance was all these two industries had going
for them, the pernicious influence they have on America's political process could be tamed as the scoundrels were eventually
named and shamed. The bigger problem is that the industries
do a brilliant job of conflating their narrow self-interests with
the larger national interest. In other words, oil and car bosses
somehow manage to persuade Americans that what's bad for
GM must be bad for America.

And they have succeeded with this tactic because the Washington policy apparatus—its historical memory and its very
soul—was formed back in an era when oil was seen not only
as plentiful but also as desirable, domestically produced, and
downright good for you. Petroleum was so plentiful at home
during the late nineteenth and early twentieth centuries that the
United States was for decades the superpower of the world oil
market. Only after World War II did Franklin Roosevelt, fearful
of dwindling domestic production, enter into a petroleum-for-
security alliance with the oil-rich new country of Saudi Arabia—thus forging an enduring Axis of Oil.

American politicians have failed to act time and again when
confronted with evidence of the dangers posed by climate
change, and it becomes clear that something bigger than mere
lobbying is at work. Cheap gasoline and the open road are such

a part of America's sense of self that these industries don't argue their case based just on narrow self-interest; they argue instead that they, like Superman, are fighting for the American way. It's no coincidence that when the elder George Bush went to the big Rio Earth Summit in 1992, the first presidential summit designed to tackle global warming, he brushed off demands for an end to his country's gas-guzzling by declaring that the American way of life is not negotiable.

Washington, D.C., has, in recent years, been trapped in a time warp. Policies are made in a fantasy world in which oil is still plentiful and trouble-free, the world always needs more roads and highways for ever more SUVs, and global warming is a harmless curiosity at worst. If you think nobody in his or her right mind could believe such things today, never mind argue such a case publicly, then you have not spent much time in Washington D.C.'s inner circles lately.

"CO_2—we call it life" was the slogan for the well-funded advertising campaign run in 2006 by the Competitive Enterprise Institute (CEI). Around the time Al Gore's climate movie was launched, ads and news reports about the ads started popping up everywhere, contesting the movie's premise. In order to achieve "balance," some news organizations felt the need to give equal time to the arguments put forward by Gore and by the CEI. Never mind that Gore's arguments were backed up by peer-reviewed science from the world's leading climate experts, while the extraordinary claims made by the CEI were held by very few serious scientists.

Worse still, poke around into the funding behind the CEI, and you find that it is merely a lobbying front for the oil and coal industries masquerading as a think tank. The shameless and unscientific campaign applauding recent increases in carbon emissions should have relegated this group to the fringes of lobbying lunacy. In fact, the CEI is one of the most influential organizations in the nation's capital these days. That is but a small example of how Washington politics suffers from the Oil Curse. But that is no reason to despair.

History shows that with enough political vision and courage, leaders can overcome the vested interests in Congress and promote sensible energy policies—though it usually takes some sort of shock to the system. America did it once before, in the wake of the painful oil crises of the 1970s. While Europe and Japan imposed energy taxes, America chose instead to regulate the auto industry through the CAFE law.

At the time, the conventional wisdom held that energy use and economic output would always grow together, but Amory Lovins had the courage to advocate the alternative "soft path." He was widely ridiculed, but the 1980s proved him right. Thanks to government policies and lingering high oil prices, the rich world's energy use and gross domestic product (GDP) decoupled, and the world's developed countries grew much more energy efficient.

The biggest success was the CAFE law. With the help of high energy prices, that law led to an astonishing increase in the average fuel efficiency of new American-made cars of over 40 percent from 1978 to 1987. From 1977 to 1985, the volume of America's net oil imports fell by nearly half, even as its economy grew by a quarter. That is proof positive that higher fuel efficiency can go hand in hand with economic prosperity—and clearly refutes the Kyoto-bashing argument that tackling global warming will inevitably lead to economic ruin. Lovins believes this "broke OPEC's pricing power for a decade" as the world enjoyed low and stable oil prices from the late 1980s through much of the 1990s.

More strikingly, he now argues that the world can repeat

that trick once again, so great are the remaining inefficiencies in energy use. After a century of refinement and countless fortunes spent on R & D, he observes that barely 1 percent of the energy in a gallon of gasoline is used to move the driver in a forward direction (which, when you think about it, is the reason anybody buys a car); all the rest of that energy, he calculates, is lost to inefficiency, grinding gears, or moving heavy metal. "I guess I see the world through *muda* spectacles," he chuckles, referring to the Japanese word for waste or utter uselessness.

Carbon Nation

The lesson from all this is that the world can do much, much better when it comes to energy—and that government policies absolutely do matter. The European Union has pushed the car industry hard to reduce carbon-dioxide emissions and showed in the middle of the decade that it was willing to enact tough legislation when the industry fell short of targets under a voluntary deal struck in 2000. Now, it is time for Washington to embrace such sensible policies. Why should it bother, ask angry oilmen, when oil has served the world economy so splendidly over the past century? Peter Robertson, Chevron's vice chairman, argues that the world simply does not appreciate the contributions of his industry to human welfare. And there is truth in what he says. The nexus of the internal-combustion engine and gasoline has indeed been an essential force behind the extraordinary economic expansion seen during the twentieth century. Even so, the environment is a force for change now confronting the oil industry, like it or not.

Concerns about oil's impact on local pollution and human health are nothing new, as anyone who has endured the smog in Los Angeles knows. And though environmentalists never advertise this fact, the oil and car industries have dealt with some of those concerns successfully through technologies such as catalytic converters and low-sulfur gasoline. Today's cars are 98 percent cleaner than those from the 1970s when considering conventional air pollution.

However, the one challenge the internal-combustion engine can never overcome is that of carbon emissions, an unavoidable side effect of burning gasoline. Daniel Yergin, a Pulitzer Prize–winning historian of oil, believes the growing popular pressure for governments to tackle global warming poses a very serious challenge to the oil and car industries. Indeed, this could be the make-or-break issue that determines their future.

Both top-down public policies and bottom-up marketplace innovations must play a role in propelling the world to the post-petroleum age of carbon-free cars. On the policy front, it looks like Washington politicians may finally be shamed into serious action by the clamor in states and cities for action on global warming, a real budding grassroots revolution. As for those marketplace innovations, will the dinosaurs of the car and oil industries finally learn to dance? Or will it be an unruly and disruptive bunch of upstarts, entrepreneurs, and innovators that leads the effort in what Amory Lovins calls a "Schumpeterian era of creative destruction"? As the next two chapters explain, the reason to think it might be the upstarts is that both the auto and oil industries in America are in bad shape.

CHAPTER TWO

Down and Out in Detroit

How the car industry, the icon of American greatness in the last century, lost its way

By the time America was waking up to the need to face its oil addiction and do something about limiting greenhouse-gas emissions, Detroit was in no shape to handle the challenge. Losses were piling up at GM, Ford, and the American end of DaimlerChrysler. Just how did the once-mighty Big Three get into such a state so quickly?

In the late 1990s, Chrysler had succumbed to Daimler-Benz, but GM and Ford were still riding high. Their profits were pumped up by a boom in demand for pickups, SUVs, and other light trucks, the part of the market where they still dominated. But events soon exposed underlying weakness. The slowdown in the economy after the bursting of the dot-com bubble and the psychological shock of September 11 to consumers changed everything. Japanese and European import brands moved into SUVs, bringing tougher competition. Incentives of up to $3,000 were needed, money on the hood, to keep the metal moving. Although sales stayed up around seventeen million in the U.S.

market, profit margins were undermined by the discounts and zero-finance deals. Detroit's apparent comeback, so lauded in the middle of the decade, went into reverse.

GM, Ford, and Chrysler (even under new ownership) were sliding into a financial mess. The dot-com bust tore great holes in their pension plans. As stocks fell, so did the value of their funds. More competition undermined the superprofits of the old oligopoly days. Too many cars were chasing too few customers. Falling payrolls, as Detroit retrenched to boost productivity, meant that fewer and fewer workers in the pension plans were carrying more retirees on their backs: by 2003, there were nearly three pensioners for every GM worker in North America. Meanwhile, Toyota, Honda, and Nissan in North America had young workers and only a few hundred pensioners, with plans less generous than those prevailing in Detroit. This was hardly an industry in great shape to face the revolution needed to cure oil addiction and cap greenhouse-gas emissions.

The problem of legacy costs goes back a long way. In the late 1940s, the United Auto Workers (UAW) wanted to have industry-wide pension funds in states such as Illinois and Michigan. Risks would be pooled across employers, and workers would get some security in their old age. But employers saw this as a dangerous concentration of power in union hands. So employers, including the car manufacturers and their parts suppliers, decided the better way was to offer company pensions to all factory workers. So the burden of today's legacy costs was created by the companies themselves. Such costs were not a problem when profits were high, but as soon as the Big Three started to lose market share in the 1990s, they had to start cutting capacity and jobs. The first problem was that union agreements meant the sacked workers had to be paid while they were on standby in a jobs bank. And those who took early retirement instead only added to the already heavy burden of legacy costs. Things could only go from bad to worse, which is why the companies had to consider help from foreign rivals.

The process that started with Chrysler falling into the hands of Daimler-Benz in 1998 was staring GM and Ford in the face by the fall of 2006. Now they were in the frame. GM's biggest shareholder, Kirk Kerkorian, insisted that the board of GM consider approaching Carlos Ghosn, who ran both Renault and Nissan in the Japanese-French car alliance. He was only too happy to consider GM joining in some form as a third leg, strengthening its position in the world's biggest car market. Bill Ford had earlier tried to lure Ghosn to run Ford. Then, as the Ford board of directors grew worried about the company's worsening position, Ford called Ghosn again to suggest that he consider hooking up with Ford instead. So rattled was Bill Ford that he called in a former Goldman Sachs banker to see which parts of Ford could be sold off so it could concentrate on its troubled core business in North America. The company had spent heavily on European brands such as Land Rover, Jaguar, and Volvo, only to find they were mostly trouble. Jaguar alone swallowed $10 billion of Ford money over seventeen years. In desperation, Ford hired Alan Mulally, the man who rescued Boeing Commercial Airplane Group in the late 1990s. Mulally's greatest challenge was reenergizing Ford's dispirited managers and keeping the Ford family from interfering.

The pace may have picked up in the past ten years, but the signs of decline and fall have been around a little longer. The Big Three have been slow to understand just how much the game has changed. As foreign competitors, still revealingly referred to as "import brands," established their own production in America, the last barriers to entry had been overcome. No longer would import tariffs on light trucks keep that part of the market safe from encroachment. The Big Three still held sway at the larger end of the SUV and pickup market, but the competition from Asian and European products made in America was getting tougher all the time.

Since the mid-1980s, when Japanese imports began to threaten them seriously, they have been marching backwards to the sound of gunfire. In the mid-1980s, Roger Smith, the chairman of GM stalked by filmmaker Michael Moore in *Roger and Me,* threw billions at robots, as if they were the magic answer. GM's factory problems stemmed from bad management and unproductive workers, and robots were no solution. Then he threw money at shareholders by spending over $10 billion buying back shares, using the company's own money, to boost the share price and appease investors. All along, he was throwing money at workers as unions forced the company to concede wage and benefit demands it would not be able to afford in the long term.

The Big Three, with labor as a fixed cost and no freedom to cut earnings in bad times (as Toyota and other Japanese manufacturers could), were caught in an infernal cycle. They could not close factories quickly enough to adjust capacity to falling demand. This made them even less competitive against the greenfield, nonunion plants the import brands opened in the southern states. The only answer seemed to be to keep production going as high as possible, to keep the cash coming in. All through the 1990s, when America enjoyed unheard-of prosperity and when the ups and downs of the economic cycle that used to convulse Detroit virtually disappeared, things went from bad to worse. All seemed plain sailing, but the sales incentives were the only source of buoyancy, while profits made on financing consumers' purchase of cars were the only way to stay profitable. Running the companies for cash and short-term survival was no match for the careful, long-term product planning and investment of their Japanese competitors.

Now, it is true that Detroit in general and Chrysler in particular have been written off many times. Every ten years or so, it seems that the Big Three are doomed. Then a few years later, the story is one of a remarkable bounce back. Just as the last rites are about to be administered, something comes along to breathe life back into the ailing body. At the end of the 1970s, a government loan guarantee saved Chrysler; in the early 1980s, the family rescued Ford; in the 1990s, SUVs turned out to be a tide that lifted all boats. Now the tide is ebbing, and the beach is looking crowded. The body has been weakened by the effects of multiple crises; its old structures and systems are not fit to cope with the complexities of the modern world. The once-proud Big Three have come to look like old people, gradually fading and shrinking in a bewildering, even alien, world.

Until recently, their sheer dominance of the American market, in the days when they accounted for two-thirds of sales, made the Big Three impregnable. The huge scale of their production facilities, of supplier chains and their dealer networks, constituted a formidable barrier to entry for newcomers. But once the import brands fanned out from their beachheads in California and on the East Coast, the game changed. Detroit's huge investments turned into barriers to exit for the incumbents. With labor-union contracts and dealer franchises limiting their ability to adapt quickly to their reduced role, their costs per car rose, making their competitive position weaker and accelerating the downward spiral they had been sucked into. Meanwhile globalization, with its reduction in import restrictions and free flows of capital, was washing away the barriers to entry behind which Detroit had long sheltered. The Big Three were stranded with decaying assets, built up in the days when together they enjoyed a tight oligopoly. The bumper profits of boom years in the long economic expansion that the United States knew postwar were artificial. They were what economists call an "economic rent" extracted from an artificial, privileged market position.

Once this was taken away by competition from imports and foreign direct investment in America, the true position of the Big Three was exposed. It was not good. The first of the Big Three to display signs of decay was Chrysler. In a way, the Chrysler story is over, because the company lost its independence and spent years of misery before its new masters got the business under control. But for Ford and GM, there is still much misery to come. If they are unlucky, Chrysler's past could be the prologue to their future. Making things even more complicated is the fact that environmental pressures are encouraging the introduction of disruptive technologies, such as increasingly electric cars, built by new entrants to the market. Detroit is still fighting a rearguard action against foes inspired and enabled by globalization, even as new entrants with revolutionary new technologies are getting ready to enter the fray.

Chrysler Crumbles

The downward spiral first caught Chrysler not once but twice before it finally succumbed to Daimler-Benz. Disaster struck at the end of the 1980s, only a decade after Chrysler had been rescued by a government loan guarantee. On February 13, 1990, Lee Iacocca, Chrysler's chairman, strode into the stark, brick-lined press rooms at the company's headquarters in Highland Park, a grimy suburb of Detroit that gave birth to Henry Ford's mass-production system. His gray, haggard face had a grim message to impart. Iacocca announced a record loss of $664 million in the last quarter of 1989 and the closure of two factories. For the second time in only ten years, Chrysler was facing bankruptcy. It had fallen to fifth place in the American car market, behind Honda and Toyota. The recovery that had started when the federal government bailed it out with loan guarantees at the end of the 1970s had fizzled out. Iacocca outlined a plan to shrink the company's capacity by a third (675,000 cars per year) with a commensurate reduction in the labor force. At first, it seemed to work.

Four years after Iacocca's solemn statement, a strange industrial ritual was taking place in a large hall at Toyota's headquarters near Nagoya, 200 miles south of Tokyo. Engineers from all over the company and from Toyota's suppliers converged on Toyota City for a *tenji kai*, an autopsy of a competitor's car. All car companies do this from time to time, tearing down a new model to uncover its manufacturing and technical tricks.

The car being taken apart this time was Chrysler's new Neon model, aimed at the cheaper end of the compact market and selling for nearly $2,000 less than rival cars from Toyota or Honda. Nissan's president, Yoshifumi Tsuji, also paid tribute to the new Chrysler model. "Where we would have five parts to make a component," he said, "the Neon has three. Where we would use five bolts, the Neon body side was designed so cleverly, it needs only three."

What the presentation revealed was a very different picture of Chrysler from the grim scene portrayed by Iacocca that somber day in February 1990. Three years later, Chrysler was making almost as much money as the then-booming Ford, with profits of $3.8 billion, half as much again as GM, even though it sold only 2.5 million cars a year, compared with GM's 7.5 million. Relocated in shiny new offices in leafy Auburn Hills (America's largest low-rise office building after the Pentagon), well away from Detroit's dirt, the reborn Chrysler was the talk of the automotive world.

Iacocca's last act before he retired from the Chrysler chairman's job at the start of 1993 was to frustrate the ambition of his number two, the company president Bob Lutz, to succeed him. Lutz and Iacocca had only one thing in common: an enormous ego. Iacocca is a straight-talking engineer who came up the hard way, while Lutz was born into a rich Swiss banking family that later relocated to America, speaks French and German fluently, and oozes elegance and charm. Even that would fail him sometimes, referring publicly, as he would, to Iacocca as a "boring old fart." Instead, the job went to Robert Eaton, who had been the head of GM's European business. Eaton quickly sold off various ancillary businesses to cut the company's debt. The core business itself was reviving, thanks to the leadership of Lutz as president. Fortunately for Eaton, Lutz stayed on. Eaton found he had inherited a company well on the way to recovery, after the cutbacks, because Lutz had put together a team of young gearheads who were passionate about cars: the opposite of the planners and accountants who had come to dominate GM and Ford. For generations, the way to make your mark and impress Henry Ford II after 1945 was to be one of the financial "whiz kids," as they were called. The same applied at GM, as the rise of Rick Wagoner to the top job showed.

Lutz was the antithesis of a financial whiz kid. He was a marketing man and a supreme gearhead, obsessed with everything automotive. Although neither a trained engineer nor a designer, he had the drafting skills to while away dull product-development meetings sketching an idea on an envelope. His drawings were good enough to inspire the experts to develop them into something real. Lutz was convinced that Chrysler could compete only if it learned some tricks from its Japanese competitors. He brought in as head of manufacturing Dennis

Pawley, who had learned Japanese lean production techniques firsthand while working for Mazda. Another Japanese trick was to break down the barriers between different departments and get engineers working as a team. His director of engineering was French-born François Castaing, who had come to the company from American Motors (bought by Chrysler from Renault as Renault began to quit the American market). Castaing formed a Japanese-style product-development process. Like his boss Lutz, he was a real gearhead; you had to edge around gleaming classic car engines in his office to get to his desk.

Castaing had grouped engineers, designers, parts buyers, and manufacturing and marketing experts into platform teams so that work on all aspects of a new model could be done simultaneously rather than more slowly, in sequence. He himself had learned the habits of teamwork and quick reactions as a young man working in the pits for Renault's Formula 1 Racing Team, when he was a design engineer. As well as its smash hits like the Dodge Ram pickup and the luxury Grand Voyager minivan, Chrysler also brought out a new range of cars that were sleeker and more nimble than the cheap old clunkers of the 1980s. Bud Leibler, Chrysler's vice president of marketing in the early 1990s, later maintained that customers in the 1980s "bought cars because Lee (whom they idolized) told them to; because they were cheaper, and, because, as the first with air bags, they were seen as safe." Chrysler customers were blue-collar and middle-aged, "somewhere between 55 and death," as Iacocca admitted later.

But the second coming of the modern Chrysler in the 1990s was more soundly based, with men like Lutz and Castaing creating a more sophisticated and nimble product-development process. In 1996, Chrysler had a return on sales of more than 6 percent, way above Ford and GM, and a return on capital of

20 percent. It also had more than $8 billion in the bank. Yet its shares traded at only just over four times earnings (a quarter of the average valuation of industrial stocks), because Wall Street feared it would lose its shirt in the next economic downturn, as it always had. The cash pile soon attracted the unwanted attention of the billionaire Kirk Kerkorian (Chrysler's biggest shareholder, with 10 percent), who said he was going to mount a $23 billion bid for the company.

He roped in the retired Iacocca in an effort to lend credibility. His game plan was to get the company to restore its cash pile to shareholders like him. He talked about bringing in Daimler-Benz as a partner. What he did not know was that Eaton had already been talking to the independent-minded Helmut Werner, who ran the Mercedes-Benz car business in the Daimler-Benz group, about a possible linkup. Desultory talk about a joint venture for markets outside Europe and North America got nowhere, but the seed had been sown. Internal studies in both companies had shown that they were ideal partners because their products and their strong markets were complementary: Chrysler was big in North America and in light trucks; Mercedes was big in Europe and in premium sedans. This underlying situation was later to spark the biggest and boldest industrial merger ever undertaken.

Changing Game Means Changing Gear

Although Mercedes was in good shape, the increasing sophistication of cars, with more and more electronics involved at the luxury end, argued for more volume over which to spread development costs. Cars are becoming ever more electric and electronic. Even as the internal-combustion engine still holds sway over alternatives such as gas-electric hybrids, battery-electric cars, or fuel-cell vehicles, it is nevertheless relying more and more on electronics to improve it. Where twenty years ago there would be one electronic engine control unit under the hood, today there are about fifty microcomputers, plus dozens of tiny electric motors to move wing mirrors, adjust headrests, regulate climate control, alter ride suspension, and so on. Drive control is the latest big area for the expansion of electronics. It is no longer just luxury models that feature sophisticated traction controls that adjust the power delivered to each of four wheels depending on the road conditions. Next to come along will be electronic systems such as brake by wire and drive by wire, whereby clunky, heavy metal units will be replaced by lighter electronic brake and steering controls.

One of the attractions of fuel-cell–electric cars, when they start to appear on the American market in coming years, will be the roomy new freedom of their interiors. With electronic controls and motors within wheels, the cabin space is freed up entirely for passenger comfort, as a whole tide of technologies rides in with the electric drive. So the revolution that will transform the auto and oil industries is already under way, under the hood of the car you are driving.

Such technical complexity adds to the development cost of new models, just at a time when fierce competition is demanding a brisker rate of product development, with new or much-improved models needed every three years. Long gone are the days when Alfred Sloan's planned obsolescence meant just some new headlights or different fins to mark the new model; now there has to be some credible technical advance. Most of these are electrical and electronic. This was the first sighting of the sort of disruptive upheaval about to upset the car industry. The quickening pace of change, the advance of electronics, the need for greater scale, the effects of global rather than regional capitalism were about to combine to change the game.

Jurgen Schrempp, by the late 1990s, was the master of the Daimler-Benz universe. He spent his first few years at the top breaking up the sprawling engineering conglomerate that the company had become under his predecessor, in everything from railway locomotives to regional aircraft, washing machines to computer software. He was determined to concentrate on the core Mercedes car business, which was still mostly in European markets, with a thin sprinkling of sales in America and around the world. The product range was very narrow; Schrempp decided to widen not just the product offering but to go for a dramatic move that would firmly implant the group in the world's biggest single car market. It would also give it the scale to cope with the disruptive technical changes and challenges that lay ahead. If new technology, such as hybrids, pure electric cars, and fuel-cell–electric vehicles were to come along and change things, the giants had to be ready: the dinosaurs must learn to dance. But first they must grow a little.

A former car mechanic, Schrempp had gone to night school and worked his way up the ranks of Germany's most prestigious engineering company by sheer ability and strength of personality. Schrempp was a gutsy, earthy, willful leader whose style was to listen before making up his mind; thereafter he dominated but did not like getting bogged down in details. He was fond of saying that when you are meeting managers in the field, the best thing is to take them out for a good meal and a bottle of wine at the end of the day. "You never get the real truth before ten o'clock," he liked to say.

After becoming chairman of the board in 1995, he lobbied his fellow board members to get Mercedes-Benz, a wholly owned but separate corporate entity, brought directly within the fold of Daimler-Benz. The only problem was that the Mercedes boss Helmut Werner, who had turned the faltering luxury

car business around after 1993, wanted to keep his operating
independence. Eventually Schrempp, a man of enormous passion and energy, wore down the opposition of all but Werner.
Mercedes was brought into the Daimler fold, and Werner left
the company, to be replaced by Jurgen Hubbert, reporting directly to Schrempp. With the organization tidied up, Schrempp
began to look at his strategic options for the group.

At Chrysler, Eaton was also looking at what lay ahead. He
saw a triple storm about to buffet the industry. First was the
overcapacity that meant the world had factories to build eighty
million cars, while sales were only around sixty million. The
Big Three in Detroit were unable to trim their capacity to match
their sales, even as the Japanese and Europeans were adding to
capacity. In Europe, Volkswagen, in which the government of
Lower Saxony was the biggest shareholder, was embarking on
a big sales expansion to make better use of its huge labor force.
Meanwhile, the French government was pouring subsidies into
a recovering Renault in order to privatize it. Japanese producers led by Nissan and Honda, followed by Toyota, were also
putting up greenfield plants in Europe in search of market share
and profits that had eluded them for years in that market. South
Korea was building up a car industry with an annual capacity
of five million cars in a market that was less than a third of
that. Union resistance and political timidity usually blocked the
closures and capacity reductions that had already come in other
mature industries, such as steel and textiles, both in America
and in Europe. This was the perfect example of an industry
where the barriers to exit were higher than the barriers to entry,
as new competitors continued to come along. The auto industry
by the mid-1990s was plainly heading for a crash.

Second, the Internet was giving the consumer information
and power that had long remained in the hands of the manu-

facturer and the dealer; customers would walk into showrooms knowing everything they needed to about the car they wanted to buy. Third, the growing pressure to develop more environmentally friendly cars was adding to the burden of development. Lutz found himself sitting next to Schrempp at a dinner at the Frankfurt Motor Show in September 1997. Schrempp asked Lutz what had gone wrong with the talks in 1995. According to *Wheels of Fire*, a racy, if hagiographic, insider account of the deal written by David Waller, Lutz said, "It was neither fish nor fowl. It was too big and too complex for a joint venture and fell well short of a merger." They agreed that Schrempp should talk to Eaton, the Chrysler boss. So started the process that would lead the following May to the shock news of the two companies joining forces.

There were inevitable culture clashes. Ironically, the man who could have bridged them because of his German Swiss–American background, Bob Lutz, chose to leave Chrysler. For a little while after him, things seemed to hum along as well as before, as the American market boomed, especially in minivans and SUVs, where Chrysler had its strength, and margins were strong. Then, in the second half of 2000, this rosy picture began to fall apart for Chrysler. Profits of $2.5 billion in the first six months were followed by losses of $500 million in the third quarter and then $1.5 billion in the fourth. The arrival of American-made SUVs from Toyota, Honda, and Nissan was destroying Chrysler's market position; it could hold on only by offering ever deeper discounts. The problem was that a market in which Chrysler had long dominated with its range of Jeeps was now peopled with more competitors.

In the middle of all this, DaimlerChrysler's share price fell from a high of $105 to only $38, worth less than Daimler alone had been worth before the merger. In a little more than two years since the merger of all manufacturing mergers, around $28 billion in shareholder value had been wiped out. Without Lutz at the wheel, Chrysler seemed to lose its way; its designs became less bold and rather staid. Lutz has been described as the Stephen Spielberg of the auto industry. He does not believe in endless market research, focus groups, and corporate criteria when it comes to evaluating proposed new models; he just has a vision of where he wants to go, and implements it boldly. His successors lost the plot, and the Daimler people back in Stuttgart, hamstrung by fear of seeming too heavy-handed, were slow to react. Had Chrysler been on its own, it really would have gone for the third time—and stayed down. But having a German parent was not sorting out its problems, until a new man was drafted in, bringing with him a trim rottweiler of an assistant.

Dieter Zetsche is a big, bald, ebullient German with a snow white walrus mustache and excellent, German-accented English. Unusually for a senior German executive, he has an impish, self-deprecating sense of humor (which was on display during the quirky advertising campaign Chrysler ran in America in 2006 featuring "Dr. Z" himself). Accompanying him in his mission to rescue the ailing Chrysler was a rising young star named Wolfgang Bernhard. Together they came up with a two-track program of big cutbacks, cost savings, and a new product strategy designed to move the Chrysler brand upmarket. Soon they had hacked 16 percent off the cost base, but by 2003, they were still losing $496 on every vehicle they sold. Six Chrysler factories were closed, and some forty thousand jobs went, but the company came way back, earning healthy profits by 2006. It even had a runaway success with its Chrysler 300, the big, bold, rear-wheel-drive sedan that is the biggest hit car (as opposed to SUV) the company has known. The Germans and Americans bedded down in recent years, after the troubled twenty-four months following the merger. The year 2005 reflected this, with operating profit in the first half of $1.3 billion, up 6 percent from the year before.

Benefits began to flow from sharing technology and hardware across the two model ranges. Indeed, as Chrysler recovered, the problem was back home in Germany, where quality problems dented Mercedes's image and sales to the point where it was overtaken by its archrival, BMW. Schrempp was pushed out in autumn 2005 by shareholders unhappy that the value of their holdings had halved on his watch. Zetsche took his place and also took direct control of Mercedes. He axed six thousand white-collar jobs and gave Mercedes the same sort of makeover he had given Chrysler. Some eight thousand production jobs were cut. No sooner was that under way when Chrysler's recovery faltered in late 2006, as stocks of unsold cars had to be written off, pushing the company back into loss. This forced Daimler to put a "for sale" sign on Chrysler; it was bought by a private-equity fund in May 2007.

The harsh lesson from the DaimlerChrysler experience is that competition in the car industry is now unforgiving: get distracted by foreign takeovers, as Chrysler did, and your competitors will eat your lunch at home. At DaimlerChrysler's competitors, such as GM and Ford, the effects of years of neglect and decline are now showing up. This is bad enough at the best of times, but with technology and environmental and energy concerns spurring disruptive future shocks, it could be a recipe for disaster. From now on, only the fittest will survive.

How Ford Flunked

The Big Two, Ford and GM, may so far have escaped the high drama that has been the story of Chrysler, but they have still lived through a tragic decline. Even for them, the drumbeat of revolution was growing ever closer, ever louder. While GM has seen its market share slide from one-third to barely a quarter, Ford has slipped from a quarter to a sixth. Now that the market is shared by six big companies, with Hyundai an aggressive seventh on the rise, there is no return to the old days. This was becoming apparent to the top echelons in Ford even as it was making record profits in the late 1990s and steadily catching up with the market leader, GM. Those were the halcyon days when Alex Trotman was the boss of Ford.

Just when things were looking bright for Ford, Trotman launched a massive reorganization called Ford 2000. It was designed to reengineer the company from top to bottom. Ford 2000 was designed to turn Ford into a truly global structure, sharing technology and designs across American, European, and Asian subsidiaries instead of reinventing the wheel in each market. Trotman reckoned that good times were ideal for instituting change. The regional baronies were to be scrapped in favor of global-platform groups for different kinds of vehicle, all but one headquartered in the Glass House, the Dearborn World Headquarters, outside Detroit.

Ensconced in the library bar of the St James hotel in the exclusive 16th arrondissement of Paris, Alex Trotman was looking back on five years as the boss of Ford. A rather stiff, reserved Scot with a clipped mustache that made him look every inch the Royal Air Force officer he had once been, Trotman made it to the top of Detroit from the poor slums on the outskirts of his native Edinburgh. A childhood acquaintance was also to become famous—Sean Connery, a milkman delivering pints to the same tenement houses where Trotman delivered meat as a butcher's delivery boy. The year before he got the top job, Ford lost a record $7 billion, and the share price was just below $12.

When he left five years later, Ford was making a profit of $7 billion, the biggest of any car company, and the shares were worth over $32. Then it was Ford, not Toyota, that was chasing GM for the top slot. But disaster was around the corner. Indeed, the $7 billion profit in Trotman's last year was followed by a loss of $5.4 billion two years later, swollen by the expensive recalls after Firestone tires were suspected of causing rollover accidents in Ford's Explorer SUV.

What went wrong? Reminiscing in Paris, even Trotman gave Ford 2000 a mixed score. It had toned up the organization by encouraging a reengineering of many business processes. But Ford also lost its edge in product development, and managers in Dearborn were cut off from what customers wanted in different markets around the world. The problem with Ford 2000 was that it was designed to solve a problem that had been self-inflicted by Ford: lousy organization with poor cooperation among rival baronies. The Detroit head office, for instance, failed to appreciate how Europe was going over to diesel engines in small and large cars, as drivers sought better gas mileage and as cleaner diesel technology came to market. The oomph went out of local marketing in national markets in Europe and South America. Ford of Europe and Ford South America had to be rebuilt to give effective decision-making power to these regions.

Meanwhile, no sooner had Trotman pushed through Ford 2000 than he was talking to Daimler-Benz about a possible merger, unaware that the German company was already talking to Chrysler. Trotman wanted to buy Detroit's number three but felt that the federal antitrust authorities would never allow it. Daimler-Benz was his second choice. The talks went nowhere, because Daimler was already speaking to Chrysler. Later, when Trotman took a call from Schrempp about the impending marriage, he hung up the phone and never spoke to the German again. Ford then went out and bought Volvo and Land Rover, which turned out disastrously.

Trotman was eventually rudely elbowed aside in 1999 by Bill Ford and the rest of the family. Bill became chairman, with Jac Nasser as chief executive. A bitter Trotman was heard to say, "Well, Prince William, you have your kingdom back." Trotman had devoted lots of time to managing the relationship with the family but felt it was a mistake not to give Nasser a free hand as chairman as well. From the start, the relationship between Nasser and the great-grandson of the founder was difficult. Within months, there were rumors of fierce disputes. Nasser drove everybody else as hard as he drove himself. He wanted to turn Ford from a stolid old manufacturing business into a consumer products and services company specializing in transport, selling not just cars but mobility. He never really delivered a business model that lived up to this slick slogan.

But that did not stop Nasser from dreaming of somehow converting the image of Ford Motor from that of a stodgy old metal basher into a purveyor of consumer services somehow related to transport and mobility. But it was largely image. He was swept along by the fever of the dot-com bubble and the promise of the Internet. His office overlooked the chimneys of the old Rouge steel plant and the original factory that was the core of Henry Ford's empire from the 1920s onward. But the interior was distinctly postindustrial. Eight television screens lined the office's back wall, above a bench with a powerful, open laptop. On a windowsill sat two more monitors showing the Web sites of Ford and its main competitors. CNBC burbled in the background; share prices and charts flickered right, left, and center on all these screens. Amid this buzz, Nasser would sit, holding an espresso from the gleaming chromed machine in a corner on a raised part of the office. Why all the screens? "I need to keep in touch," he grinned.

Nasser drove Ford Motor into parts of the automotive services business outside stagnant car markets in pursuit of growth and higher margins. By doing this, he sought better returns in parts of the transport equipment and services sector beyond cars themselves. This was actually quite a smart, down-to-earth hunt for value. But it had to be done at the right price. Jac was in too much of a hurry. He overpaid for the British-based European muffler-fitting chain, Kwik-Fit, which was later sold at a huge loss. He got Ford out of the original equipment parts business, selling Visteon even as he bought a chain of scrap yards. He tried to buy up chains of Ford dealers in North America, making enemies of a vital constituency that the Ford family was close to. He brought in tougher annual appraisals for managers along with a rule that the bottom 10 percent of them be asked to leave the company. Such rigorous approaches may work well in hard-driving companies such as McKinsey or General Electric, but it was a direct challenge to the cozy atmosphere that prevailed in the managerial ranks of Ford.

Nasser's vision may have been ahead of its time, but he had the bad luck to take the chief executive chair just when the problems created by the botched Ford 2000 reorganization were beginning to kick in. Losses appeared in Brazil and Argentina, while Ford Europe went into a tailspin. Ford Europe, which had previously sustained North America in bad times, started losing about $500 million a year. With competition heating up in the SUV and light-truck sector in North America as transplant Japanese factories came onstream making these models, Ford could no longer look forward to fat profit margins on such products. The weaknesses in Ford's products, quality, and manufacturing efficiency began to appear. This was no situation for taking your eye off the ball. But that is precisely what the busy Nasser did as he pursued his automotive vision over the horizon. While he was buzzing around setting up new Internet businesses, buying other companies, and chasing value up and down the automotive supply chain, Nasser was neglecting the basics of the business. Quality began to slip and push up warranty expenses to fix faults; product launches were rushed and mishandled. The tough, driven Aussie was overstressing a rather staid, traditional company with a paternalistic culture.

When the Firestone tire scandal broke in the summer of 2001, Nasser fumbled. He made TV ads telling consumers there was no problem with Ford vehicles and at first refused to testify before a congressional committee examining the growing consumer scare. Above all, he annoyed the Firestone family, which was married into the Ford clan. Having upset Ford's managers, its dealers, and its customers, Nasser finally had fallen out with the Ford family: Bill Ford's mother is a Firestone. Bill Ford got board backing to sack Nasser. Ford himself took the chief executive job, flanked by two British veterans, Sir Nick Scheele and David Thursfield, an unlikely combination of two individuals who could not stand the sight of each other, much less work in harmony.

Since January 2002, Ford has undergone several rounds of cutbacks, closures, and job losses. It never seems enough to stop the losses. When a plan launched in early 2006 failed, Bill Ford turned to an outsider to run the company. Alan Mulally, the former Boeing boss brought in to run Ford in mid-2006, was the outsider—an aviator brought in to arrest Ford's freefall. Mulally's great achievement was to get a staid, conservative company like Boeing to adopt a radically new business model and innovative technology to recover from losses and falling market share in the late 1990s.

He brought in risk-sharing partners and outsourced over a third of Boeing's new planes' manufacture to them. He sold factories making big chunks of Boeing fuselages to private equity firms and then bought the parts from them at much-reduced prices, once the new owners had cut costs, wages, and head count. Most of all, he got Boeing to make its new aircraft, such as the smash-hit 787, from lightweight plastic composites rather than aluminum. The man who saved Boeing will no doubt apply much-needed fresh thinking at Ford, but whether that will be enough to return Henry Ford's creation to the forefront of innovations (as the company's ubiquitous advertisements claimed in 2006) remains to be seen.

The General's Decline and Fall

The years since the late 1990s should have been, on the face of it, the most benign environment Detroit had ever seen. The economy is prospering, and car and light-truck sales have stayed buoyant. Yet Detroit is on its knees. The best way to explain this paradox is to look at the huge addition to America's carmaking capacity since the mid-1990s. It went up from fifteen million cars and trucks a year to over nineteen million, mostly the result of the opening of transplant factories by import brands. The Japanese Big Three, Hyundai, BMW, Mercedes, and Volkswagen now make over three million cars a year in North America. Throughout the 1990s, the Japanese manufacturers alone accounted for one-third of the $3.1 billion investment in America's car industry. This came on top of an earlier wave of foreign direct investment in the 1980s worth some $3.4 billion. States in the southern part of America have fought each other to land these investments, offering big tax incentives in return for the jobs the newcomers provide. Add in imports of some two million cars a year, mostly from Japan and Southeast Asia, and you get overcapacity in the order of 25 percent.

In the aftermath of September 11, Americans did the unthinkable; they stopped even going to car showrooms for a few days. In the deathly hush that followed, GM launched a new program of incentives called "Keep America Rolling." This was a quick and pragmatic way to jolt the American economy into life again and prevent paralysis by stimulating the nation's largest manufacturing industry and sales of the very emblem of the American dream. The move was widely praised in political and economic circles, from the White House downward, and it certainly succeeded in getting the metal moving again, avoiding a pause that could have tipped a shocked American nation into a consumption-led recession. But as a side effect, it raised the stakes in the game of discounting, which had been becoming more widespread over the previous five years. Immediately, Ford and Chrysler were obliged to follow in a suicidal game of chicken: soon incentives were worth about $3,000 per vehicle. Significantly, the Japanese, Asian, and European import brands were managing to sell their cars while offering discounts less than half those of the Big Three. This was yet another reflection of the fact that the "import" brands were selling on production and reputation, while Detroit still competed largely on price. The Japanese and Europeans sold cars; the Big Three sold deals.

Against this background, GM has been as slow as the other Detroit companies to downsize, to adjust its capacity to the number of cars it can sell profitably. It took out about one million units of capacity between 2003 and 2006 and planned to eliminate another one million by the end of 2009. The trouble is that even when a factory is shut, the company has to put the redundant workers into a jobs bank, where they draw most of their wages in return for showing up for "training" every so often. This is part of the union agreements that Ford and GM have long had with the UAW. It took until 2006 for GM to make an agreement with the unions to buy out over thirty thousand jobs to get itself into shape.

There have been occasional false dawns amid all this tension. Cadillac came back from the dead as a luxury brand and even reentered the European market, taking on the local champions, BMW and Mercedes. After Rick Wagoner brought in Bob Lutz as number two in 2001, things seemed to look up. Wagoner streamlined the corporate bureaucracy, and Lutz shook up the cumbersome product-development process. For a man in his midseventies, Lutz cut an impressive figure as he returned to the stage at the Detroit Auto Show, driving the prototype of a new roadster. He promised "turmoil and change." He certainly brought more pizzazz. A man who flies his own helicopter to work every morning, he spends Sunday afternoons streaking through the skies of Michigan in his old Czechoslovakian fighter plane, as if to relive his days as a pilot in the U.S. Marine Corps. One Detroit show featured a video of him landing the plane with a nervous-looking Rick Wagoner seated right behind him, in the navigator's seat. Such was the competition that GM needed superior design and enticing products, but it takes up to four years to refresh a weak product range, so nothing was possible overnight.

It is possible to feel sorry for the Big Three of Detroit. They have, after all, made remarkable progress in improving much of their performance, especially in terms of quality and efficiency. Industry surveys have shown Detroit steadily catching up with the Asian competition in terms of productivity. In 1998, for instance, GM and Chrysler each took about forty-seven hours to build a whole car, while Toyota took thirty hours. By 2006, the

American pair had cut the total hours to thirty-three or thirty-four hours, while Toyota had only just nudged below thirty hours. But the Japanese had profits of more than $1,200 per vehicle, while GM lost $2,496 per vehicle, and Ford lost $590. Chrysler made a small $223 profit per car. The difference between these profits and losses is mostly the legacy pension and health-care costs borne by Detroit, compared with the Asian and Japanese producers, who employ nonunion labor and have a younger workforce. The transplant factories all work flat out, while GM and Ford by middecade were still carrying some plants that were half empty. The industry of industries was, like the oil industry, afflicted by the curse of plentiful petroleum in the first half of the twentieth century. Once oil security and the environment became a problem in a globalizing world, Detroit was in no position to make an easy adjustment.

Besides improving the efficiency of their factories and adjusting capacity, GM and Ford still face problems with consumers. The quality of their cars has improved, in terms of both buyers' surveys and professional audits of cars rolling off the line: Ford in particular has come up a lot, with a number of cars and plants featuring among the most reliable. But consumers still remember the early years of the twenty-first century when Detroit still lagged, and it takes time to regain the confidence of car buyers long used to the exceptional quality and reliability of their Camrys and Civics. Detroit's cars may often still be bland, but they are of much better quality than ever before.

The future for both GM and Ford lies in imitating Chrysler: getting smaller and smarter. The Chrysler 300 and its cousin, the Dodge Magnum station wagon, not only did a roaring trade, they regilded the whole image of the company at least for a while. But even the rejuvenated Chrysler began to suffer as higher gas prices drove customers to smaller vehicles, while profit margins kept being squeezed by the rising costs of raw materials such as steel and plastics. GM and Ford have to do more than shrink their capacity. They need to winnow their brands and models to concentrate on putting resources behind winners. Nostalgic commentators harp back half a century to the days when GM ruled the roost and sold over a million Chevrolet Impalas a year. The easy rule of the Detroit oligopoly has gone the way of the hula hoop. The American market is fought over now by a Big Six rather than a Big Three. Yet Detroit still behaves as if nothing has changed: it keeps its same bloated structures. It is worth noting that Toyota has only four brands around the world, while GM at one point had no fewer than fifteen, including associates. Toyota may stick to core brands, but it sprays the market with models, no fewer than sixty in Japan alone, to cope with the fragmentation that has hit the car market, as it does all mature markets for consumer goods.

GM and Ford could yet topple over the edge into bankruptcy. But it is more likely that they will keep going by shrinking, by paying off surplus workers, and by forming alliances with foreign automakers (or even with each other). Ford's decision to stop paying dividends is one symptom of how severe the crisis there has become. No wonder Ford family members often muttered about the harsh, unfair judgments of Wall Street. Talk about taking the company private started to be heard at the end of 2005, as the extended family got together over Christmas reunions.

The relationship between the auto industry and the financial markets has changed over the years. Nowadays, the auto industry represents barely .5 percent of America's stock market valuation; two decades earlier, it accounted for 4 percent. So this mature industry has all but disappeared from the equity market. It now relies on the bond market, and even there, much of the capital raised is gotten by the finance arms of GM and Ford. They have long since become consumer-lending businesses that have cornered the market in financing the purchases of the biggest consumer durable of all, the automobile. But now the financial tail wags the industrial dog.

It is also justified to criticize the Big Three for spending billions on dividend payouts and on share buybacks for the benefit of stockholders. They were doing this from the mid-1980s onward, when the companies could have used the money to improve their underlying industrial and business performance rather than indulge in financial engineering. But, facing increased competition in an overall market that has matured, there came a point when financial engineering had to take precedence. The only way to keep shareholders happy was to keep up that stream of dividends paid out of the profits made by lending money to people to buy their cars. They had long since lost faith in any capital appreciation, as more exciting investment opportunities came along, and prospects just got worse and worse for the incumbent Big Three.

Compare the situation of GM and Ford with that of Toyota, the world leader. In 2006, it was worth more than the Big Three put together and more than its two Japanese rivals. It was just about the only volume producer to make enough profit to exceed the cost of its capital, over time, though Nissan can challenge it on that count. And most important, it was mopping up about 50 percent of the growth in the world car market. Toyota does not have to rely purely on its consumer-finance business to make a profit: it has perfected the business model of making cars profitably, one that others struggle to emulate. When you look beyond the shores of the United States, with its rather mature market for seventeen million or so vehicles a year fought over by six leading producers, there is a wider global prospect.

Toyota's aim is to have 15 percent of it by 2010. Put bluntly, Toyota wants nothing less than to be the global number one, surpassing the venerable GM. It reached the number-one slot in the first quarter of 2007.

Meanwhile, GM and Ford's weakness means they are hobbled in the race to develop the cars of the future. GM used to make a great song and dance about the fuel-cell vehicles it was developing but seems to have gone rather more quiet on that front. Its cooperation deal for fuel cells with Toyota has been abandoned. There was a dispute over intellectual property rights that GM wanted to retain, while the Japanese thought they should be made available to all comers. That says much about Toyota's confidence in its ability to compete and about its real commitment to dealing with environmental concerns.

It sees challenges as market opportunities. Toyota's strong position means it has the power to capture much of the market growth in places such as China and India, as well as to take bold, leading positions in developing the cars of the future. Even as they publicly complained about the limitations of the gas-electric Prius and muttered about how much money Toyota must be losing on each one sold, Detroit executives knew that they would love to be in the position to do what Toyota is doing. To complain that the world's leading car company with its huge profits is losing money while launching a new model while it was spending no less than 5 percent of its revenues on developing such innovative vehicles is simply to show how far behind the game Detroit is. In fact, Toyota and Honda, the leaders in hybrid gasoline-electric cars, are long past the point where they began to make profits from their technical innovation, all the while paving the way for a world in which the internal-combustion engine and rule of the traditional auto will be challenged.

The big question now is this: can Detroit ever recover

enough to keep up with these new challenges? Can it reinvent itself with a new business model, or at least copy elements of Toyota's strategy, or is it doomed to a painful descent into the trash can of history? As Lee Iacocca himself put it in the TV ads for Chrysler in the 1990s, "Lead, follow, or get out of the way." Such questions may be painful to pose and difficult to answer for the automakers, but the opposite should be the case for the oil industry.

America's booming oil majors look like the other side of the Detroit coin. They make profits big enough to dwarf even Detroit's losses. Yet Big Oil is in a position in the first decade of the new century that resembles the position Detroit was in during the last ten years of the twentieth century. On the surface, things are looking good. But underneath, there are huge problems gushing up.

CHAPTER THREE

Big Oil in Big Trouble

The world is not running out of oil, but America's oil giants are in trouble even so

One otherwise ordinary day in February 2005, something extraordinary happened in the oil patch. Buoyed by news that ExxonMobil had earned a whopping $25 billion in profits the previous year, investors sent the company's market capitalization shooting past $400 billion. The stodgy "Old Economy" company created by the merger of two descendants of John D. Rockefeller's original Standard Oil became the most valuable company in the world, topping such icons as Wal-Mart, Microsoft, and GE.

The man who led Exxon to the top of the heap was Lee Raymond, the company's boss for much of the past two decades. The first thing to strike any visitor to his luxurious office in Exxon's headquarters near Dallas was the large portrait of a handsome Raymond hanging on the wall—surprising, given that the man standing beneath the painting was rather less attractive. The second thing to strike a visitor was surely Raymond's blunt talk. He was unabashedly dictatorial: "Everyone at Exxon works for the general good," he would roar, and add, pointing to himself, "and I'm the general of the general good." And he was a stout defender of the oil economy. Renewable energy was "a complete waste of money," he would insist, and global warming merely a government "hoax."

Though his abrasive style and unorthodox scientific views led to angry activist boycotts of Exxon, Lee Raymond was much admired within his company. He was such a hero, in fact, that when he finally retired, at the end of 2005, the board of directors sent him off with a retirement package worth a reported $400 million. That payoff was so large, it caused a popular backlash. Americans had already been suffering from a steady rise in gasoline prices after Hurricane Katrina hit in 2005, with parts of the country paying $3 a gallon or more. When one oil major after another reported record profits, the public grew suspicious. Newspapers and networks began attacking Big Oil for price gouging.

When Raymond's pay package hit the news, the outrage could simply not be contained. Bill Maher, a political satirist on the HBO television network, captured the mood with a running gag comparing Raymond to Fat Bastard, a grotesquely obese and greedy character from the Austin Powers flicks. Unfortunately for Raymond, Maher got hold of an undoctored portrait of the oilman in which he looked rather like the movie villain. So sour was the public's mood that Congress started to bash the industry too. Politicians demanded investigations into price gouging and threatened to slap the majors with a windfall profits tax. Even Republican Senator Pete Domenici, a longtime friend of the energy industry, declared angrily, "Oil companies have failed to tell us and show us what they are doing with these profits that justify them." He and his colleagues ordered the bosses of big petroleum companies to come to Washington,

D.C., and they put on the biggest show trial since the bosses of Big Tobacco firms were similarly grilled in front of television cameras a few years earlier. The ritual attacks on Big Oil during 2005 and 2006 proved mere political theater. Exxon was not shamed into cutting Raymond's payoff, Congress imposed no windfall taxes on oil profits, and no oilmen were penalized for price gouging.

Given all that, it might very well seem to the casual observer that the oil industry is still riding high. Look closer, though, and you will find that this popular interpretation of events is fundamentally wrong. There is indeed something rotten in the state of Big Oil, but it isn't quite what you think. The obscene payoffs for oilmen like Raymond and the record profitability of Western oil companies seen in recent years appear to be signs of an industry in suspiciously good health. In reality, the oil industry faces a bleak future.

It is a relic from the early days of industrialism, a century-old dinosaur clinging on desperately to the old Old Economy. Longer term, the industry faces the twin threats of geopolitics and global warming. Just as the American car industry, once the world-beater, now trails the Japanese in dealing with the challenges of the future, so too does the once-dominant American oil industry trail the European oil giants in dealing with these two threats.

However, even before tackling those longer-term threats, the industry must deal with a clear and present danger today: replacing reserves. There are signs that oil companies are actually having a difficult time replacing the petroleum reserves they pump and sell every day. That, argue some, is an early warning sign that the world as a whole is starting to run out of oil altogether. If that were true, then the world economy itself and not just Big Oil would really be in big trouble. But is it true?

The Bottomless Beer Mug

The leading lights among the petro-pessimist camp are Colin Campbell and Jean Laherrère. In a celebrated article published in *Scientific American* in 1998, they predicted that the global peak of oil production would be reached in 2004. Those who think this day of reckoning will come sooner rather than later usually point with reverence to the so-called Hubbert's peak. M. King Hubbert was a geologist at Shell who correctly predicted in 1956 that America's oil production would peak and begin to decline around 1970. The heart of the current controversy over depletion is the question of when the global Hubbert's peak will be reached.

The end is near, argues a flood of recent books with such titles as *Out of Gas* (by the California Institute of Technology's vice provost, David Goodstein) and *The Empty Tank* (by Jeremy Leggett, a British clean-energy pioneer). Kenneth Deffeyes, a respected professor at Princeton who has actually worked with the legendary Hubbert at Shell's research labs in Houston, also made a splash with his book *Hubbert's Peak: The Impending World Oil Shortage*. And Matt Simmons, a prominent invest-ment banker, in *Twilight in the Desert* even questioned whether Saudi Arabia, the kingpin of oil, has a future as a reliable oil supplier.

A principal argument put forward by this camp is that there is a fixed amount of oil in the ground to be found, and that mankind has already found it. As one depletionist says, "Dis-covery clearly peaked in the 1960s. We are out of North Seas." He argues that annual oil consumption has exceeded new dis-coveries since the 1980s, suggesting that the world is running down its earlier stock of "found" oil. He reckons eighteen ma-jor oil-producing countries, currently making up about 30 per-cent of world output, are now past their peak. Given that oil

companies have poked and prodded all the earth (save Antarctica) for over a century, goes the argument, surely there are
no more "supergiant" fields like Saudi Arabia's Ghawar, which
alone produces a whopping five million barrels per day. Campbell has neatly summarized this view of the problem: "Under-
standing depletion is simple. Think of an Irish pub. The glass
starts full and ends empty. There are only so many more drinks
to closing time. It's the same with oil. We have to find the bar
before we can drink what's in it."

Hang on a minute, though. Doesn't that ignore the barrels
in the basement and at the brewery and behind it in the rolling
fields of barley and hops? To find out, one of the authors spent
a long weekend cloistered away with the Peak-Oil thinkers at
Ankelohe, a farm two hours' drive from Hamburg, Germany.
Leading experts on oil depletion—from Colin Campbell and
Kenneth Deffeyes to Jeremy Leggett and David Goodstein—flew
to this bucolic but remote location for a conversation about energy and climate change. The Kleveman family had converted
its ancestral holding into a comfortable if modest conference
center, with the sheep stable divided up into bedrooms, and the
barn serving as the main dining hall. The sons had maintained
the horse stables and the bonfire pit in the woods that is a fine
northern European tradition.

The most important feature of this venue, noted one partici-
pant, was that Ankelohe was "a remote place where nobody
can run away once you are here!" She was not kidding. Without easy transport, limited Internet and cell phone access, and
no urban diversions at hand, the gathered crowd was forced to
focus on the task at hand for two and a half days.

The plan worked brilliantly, and it revealed something quite
important about the petro-pessimists, whom some people dismiss as lunatics. First of all, many of them are decent old men

of the oil and gas industry who, near the end of their professional lives, have decided to warn the world of this particular
danger arising from oil addiction out of a sense of duty (and,
perhaps, guilt for having peddled petroleum all their lives).
Spend a weekend with them, and you feel as though you have
spent a lifetime with the lead character in Eugene O'Neill's *The
Iceman Cometh*. In the drama, a reformed alcoholic returns to
his old bar to beg, badger, and bully his former drinking bud-
dies to kick the habit.

So it was rather cheeky of Lutz Kleveman to introduce the
depletionists with these words: "If we're to assume oil is an addiction, these distinguished gentlemen are ex-pushers who've
decided to blow the whistle." Everyone smiled, but it was clear
that there really is something messianic about their mission.
Campbell argued at Ankelohe that Peak Oil will mean the inevitable "end of the U.S. empire," the start of resource wars,
and the arrival of the second Great Depression. Deffeyes insisted that "Ford and GM are on the last few swirls on the toilet
bowl before getting flushed." Leggett said, "I really hope I'm
wrong, but I can't see any way around major economic disloca-
tion due to the looming energy crisis." Goodstein added the climate dimension to the worries: "We are doing an uncontrolled
experiment on the only planet we have. It's a very foolish thing
to do." Things only got gloomier from there as these earnest,
smart, and passionate men went on to explain why they believed the end was nigh.

Given the excess of pessimism, it was tempting to dismiss
their message out of hand. Indeed, when Colin Campbell returned for yet another refill from the fireside bar at the bonfire
pit that evening, a conference attendee could not help but nee-
dle him: "You know, Colin, there is no such thing as a bottom-
less glass of beer." He laughed good-naturedly. Jeremy Leggett

was on hand to hear the exchange, and he would not let the jibe go unchallenged. "Whenever oil peaks, the real question is whether the dam will break in favor of renewable fuels—or in favor of carbon-intensive fuels made from tar sands, shale, and coal."

That sharp insight points to the second reason to pay attention to the depletionists, alarmist though they may seem. If they turn out to be right, the two great challenges posed by oil (geopolitics and climate change) would be made even more painful. If conventional oil were really to peak suddenly and soon, then the sort of economic shock and resource wars foreseen by Campbell might be a possibility. And if the world deals with Peak Oil by moving rapidly to gasoline made from the vast but incredibly dirty "unconventional" oil deposits (such as Canada's tar sands and America's shale), then all hope of tackling global warming will be lost. That is because the carbon contained in the world's conventional oil pales when compared to the carbon contained in unconventional hydrocarbons. Burning all that mucky unconventional oil would, as Leggett points out, make for a climate nightmare.

Taking the depletionists seriously doesn't mean one has to stop the world. In fact, suddenly halting oil use would definitely lead to a Great Depression today, depriving the world of the prosperity and future technologies needed to solve this difficult problem. Instead of panicking, the wise course is to work for a sensible set of public policies that will speed us to a carbon-constrained world beyond petroleum without pushing us into the Dark Ages in the process. And surprising as it may seem, the Big Oil companies—whom you might expect to fight oil alternatives tooth and nail—are now actually investing in a variety of alternatives to oil. Why? Out of sheer desperation: the world as a whole is not close to running out of oil, it turns out, but Big

Oil might be. That shows the limits of the pessimistic argument put forth by Campbell and his peers and points to why the good times may not last forever at oil companies.

Big Oil in Big Trouble

Given the oil price boom of the early 2000s, it might seem that the oil industry is entering an era of extraordinary and enduring profits. In fact, quite the opposite is true. Those eye-popping profits disguised a fundamental rot at the heart of the industry, one that could even lead to the disappearance of the private-sector oil companies altogether—unless the dinosaurs learn how to dance.

This must seem an incredible assertion. After all, it is not just Exxon that has enjoyed good fortune of late. It seems all of the oil majors have been enjoying a golden age. Royal Dutch/Shell, shrugging off an accounting scandal involving the misreporting of oil reserves, posted the highest annual profits in British corporate history in early 2006. On both sides of the Atlantic, oil majors have so much cash in hand, they have been shoveling it back to shareholders as share buybacks, special dividends, and the like. Douglas Terreson of Morgan Stanley, an investment bank, who correctly predicted the wave of mergers in the late 1990s that created such supermajors as ExxonMobil and Chevron Texaco, declared in 2005 that "the industry is the healthiest it has ever been."

Is that really true? Even allowing for the softening of oil prices in 2006, the profits this decade have certainly been breathtaking. Industry boosters argue that handing back cash to shareholders shows extraordinary capital discipline. After all, during previous booms, managers splashed out on gold-plated acquisitions. Today, ExxonMobil runs its postmerger empire with only about as many employees as it needed to run just Exxon's assets a few years ago. The result is that the majors now deliver excellent shareholder value. The capital discipline is laudable but does not guarantee a bright future for the majors. For one thing, the chief factor behind today's profits was the surge in oil price seen earlier this decade. And look beyond that mountain of profits, and you find that the industry faces several serious challenges that could ultimately wipe out some or most of these firms, once venerated as the Seven Sisters. Just as that airbrushed portrait of Lee Raymond conveyed a too-rosy impression of the real thing, so too does today's mountain of profits convey a false sense of petro-prosperity.

That is because the world's oil majors are starting to run out of oil. To be more precise, they are finding it ever harder to replace the oil that they withdraw from their tally of booked reserves and sell every day with new reserves of the gooey gold. And even their mountain of profits has not solved this problem. That's because the biggest firms are getting locked out of the most promising oil prospects, especially those vast, untapped resources of the Persian Gulf. Oilmen vigorously deny it, but it seems the majors are cash-rich but opportunity poor—just at a time when their dwindling reserve base needs serious replenishment.

Running to Stand Still

"Oil is a depleting asset. Every day if we don't spend money and find more oil, we lose assets. Most oil companies, by doing nothing, will shrink to one-fifth today's size." So declared Steve Farris. He is the president of Apache, an independent American oil exploration firm. He was not exaggerating, for oil fields start to lose reservoir pressure and decline in myriad other ways the moment companies start pumping the black gold. In some places, like Venezuela, the annual depletion rates can be 25 percent or higher. Lord Browne, chief executive of BP until 2007, put it this way: "Running an oil company is like running up a down escalator all the time."

That points to the biggest immediate threat confronting the majors today—the reserve challenge. According to the International Energy Agency (IEA), the world will need to spend $3 trillion over the next twenty-five years if anticipated global oil demand is to be met. Most of that money will go not to increase net global supply but merely to replace output from today's aging fields. The industry has always been "self-liquidating" in this way, but the mergers of the late 1990s magnified the problem, since each company's reserve base is now much bigger.

This reserve challenge is made worse by several other factors. First, much of Big Oil's production today comes from large fields in places like Alaska, the Gulf of Mexico, and the North Sea. These reserves represent the first great wave of noncartel exploration. These fields saved the Western majors after they got kicked out of OPEC countries during the nationalizations of the 1970s. Crucially, the output from these friendly fields proved a check on the cartel's market power. Now these fields are entering a phase of rapid decline. Companies are spending ever greater amounts on fancy technologies and "enhanced" oil recovery techniques. Field maintenance costs are soaring.

The troubles in North America and the North Atlantic have sent the majors scrambling to far riskier oil provinces in developing countries. Western firms are now betting their futures on growth in such far-flung frontier regions as the Caspian, West Africa, and the ultradeep waters off Brazil. Their biggest hopes have been pinned on Russia, which opened up to private investment in oil under Boris Yeltsin and saw an extraordinary surge in investment and production.

Unfortunately, this new wave of oil exploration looks to

be rather trickier than the first. For a start, it usually involves technically complex oil formations that require lots of high technology and up-front capital expenditure. This is leading to rising costs. Another problem is that the legal regimes in these countries are hardly as reliable as Britain's or America's. For example, emboldened by high oil prices and Hugo Chavez's nationalism, Venezuela has tried to change existing contracts. Exxon's blunt retort was that "we as a company still believe in the rule of law and the sanctity of contracts." Exxon took a hard line with Chavez, but all of the other majors there swallowed their pride and agreed to the new terms. That is because they are painfully aware that the playing field of international oil is increasingly tilted toward the home teams.

The Age of Asymmetric Warfare

Fu Chengyu seems the very model of a modern major's general. The American-educated president of the Chinese National Offshore Oil Corporation (CNOOC), a partially privatized energy firm, likes to hold early-morning meetings at his modern headquarters in Beijing. He peppers discussions of corporate strategy with vows of "shareholder value" and "healthy returns on capital." His chief financial officer boasts a finance degree from the Massachusetts Institute of Technology.

Do not be misled by his Western ways. Fu is a proud Chinese nationalist and believes that his country's resources are best cultivated by local companies like his. "We developed our domestic oil industry over the last thirty years without the majors," he boasts. He also rejects the argument put forward by the majors that developing countries need them for access to fancy technology and capital with a cruel dig: "Technology I can get. Money I have. But if you don't have reserves and production, nobody can help you!" Such swagger used to be limited to a small handful of nationalized oil companies (NOCs) in the Middle East, especially Saudi Aramco, but no longer. Indian and Chinese government oil executives have spent billions of dollars on a global scramble for oil and gas to feed their booming economies. Behind Russia's bumbling crackdown on Yukos lay an audacious plan to turn the state-run Gazprom into a national oil and gas champion. In Venezuela, Hugo Chavez installed political allies in key management jobs at PDVSA, the state oil monopoly, and has talked of forging a pan–Latin American "Bolivarian" oil company.

Why is resource nationalism on the rise now, exactly? One explanation usually offered is "energy security," a woolly and much-abused notion. Post–September 11, goes the argument, the energy world is much riskier than it was during the go-go 1990s, when governments were largely content to let open markets and global trade deal with matching up supply and demand. Now, argue thinkers in big, consuming economies like China, countries need to lock in "equity oil" to have peace of mind. As for countries with lots of hydrocarbons, they are increasingly clamping down on foreign or private investment. Amy Jaffe of Rice University, who has studied the oil nationalizations done several decades ago by Iraq's Ahmed Hassan al-Bakr and Iran's Mohammed Mossadegh, argues that those earlier takeovers were driven as much by economics as ideology. She worries that today's nationalism in places such as Venezuela and Russia could wind up even more ideologically driven. That may be so, but there is just a chance that today's high-minded nationalists are merely opportunists taking advantage of high oil prices. One Chinese expert insists companies heading overseas are just empire-building. Similarly, one Russian oil oligarch argues that "perhaps some officials really want a national oil company, but this is mostly guys enriching themselves, wrapped in the national flag."

Whatever the cause, the impact of today's government intrusions into the oil business is likely to be bad news for ordinary consumers everywhere. It seems odd to say it, but the citizens of countries blessed with oil may be the biggest losers from the rise in oil nationalism. History suggests that countries in which governments play a strong role in developing oil resources tend to end up with corrupt, inefficient energy firms. There are exceptions, of course: Malaysia's Petronas is a well-run company, and such developed countries as Norway and Canada have managed their oil wealth pretty well. On the whole, though, the oil bounty tends to get misspent, and the poorest rarely see the benefits—a phenomenon known as the Oil Curse. Despite the hundreds of billions of dollars earned by their countries from oil, average Venezuelans were actually poorer in 2000 than they were thirty years earlier. Nigeria is another notorious case of oil-fired corruption.

A more obvious loser is the global energy consumer, who may have to endure higher prices over the long term if the NOCs in OPEC—or those from countries like Mexico and Norway that collude with the cartel—increase market share at the expense of the majors. And the cartel's coffers are once again overflowing. Until that happy day when petroleum is irrelevant to the world economy, a robust private-sector oil industry and its non-OPEC production will remain an essential counterweight to OPEC, especially since the world's remaining reserves are highly concentrated in the Middle East.

The Anti-OPEC?

So it is a pity, then, that the biggest loser from the rise of resource nationalism looks to be Big Oil. NOCs are increasingly doing battle with the majors outside their home turf, but doing so with unfair advantages arising from their quasi-governmental status. One clever oilman calls this the coming age of "asymmetric warfare." That is defense-industry jargon for an unfair war launched by a stealthy, guerrilla-style adversary on a big, easily targeted foe—the economic equivalent, one controversial thinker argues, of al-Qaeda's cowardly attacks on American military and diplomatic targets.

One example is oil diplomacy. One of the Chinese NOCs recently struck an oil deal in Venezuela on the heels of a diplomatic warming between the two countries. "Who knows whether that deal was at commercial terms?" grumbled one private-sector oil executive. Indian officials talk breathlessly of a race with China to lock in energy assets and brag of mobilizing a cadre of ex–foreign service officers to schmooze for Indian NOCs. The generosity of the Chinese government in building railroads, ports, and other bits of infrastructure in Africa has greased the skids for Chinese oil and mining companies keen on gobbling up local assets. Because state-controlled firms rarely have to meet the same standards of transparency as publicly listed ones, nobody knows their true financial status. Most operate with "soft" budgets and know that their state parents will supply any needed capital. Anecdotal evidence suggests that government-run firms are overpaying for assets, at times, in order to edge out majors, because they rarely need to meet rigorous tests of returns on capital employed.

Another edge NOCs sometimes exploit is that they worry less about environmental and human-rights campaigners. When activists forced Canada's Talisman Energy to stop doing business with Sudan's thuggish government, the oil did not stop flowing. Indian and Chinese firms happily stepped into the breach. One private-sector oil executive points to a special supplement on Indian oil carried by the *Financial Times* and cries in exasperation, "See this, they are bragging about operations in Myanmar. No major could touch that project!" Some human-rights cam-

paigners even see China's oil interests as a big obstacle to the
UN voting to intervene and stop the massacre in Darfur: China,
a Security Council member, would not want to upset its petro-
partner in Sudan.

Even in countries that do not have a lot of reserves, the locals increasingly have the upper hand. The majors have been
salivating over the chance to peddle gasoline to China's billion-
plus consumers, a growing number of whom will be driving
cars in coming years. However, the bonanza may not extend
to the majors. One ominous sign was the fiasco involving a
lengthy gas pipeline from China's far west to Shanghai. Exxon
and Shell had hoped for access to the Chinese market, including
a lucrative slice of the upstream gas reserves, in return for helping build the pipeline. In the end, the terms proved so unattract-
ive that they all pulled out; state-run PetroChina completed the
job by itself.

Majors had also hoped to crack China's vast retail market
to plug their global brands. Firms like BP have indeed built a
few stations with local partners. Argues David Hurd of Deutsche
Bank, an investment bank, "China has 1.3 billion consumers,
just like Russia has vast quantities of oil and gas, but neither is
going to give them up to foreigners. China wants technology, but
then will spit them out." The Russian "oiligarch" concurs, arguing that for foreign firms eager to buy Siberian oil assets, "the
window of opportunity is closed now."

The Coming Shakeout

So will the rise of NOCs endanger what's left of the once-
proud Seven Sisters? The challenges facing the majors are indeed formidable: rising costs, declining reserves, unreliable
prices, lack of access to reserves, and now unfair competition
from subsidized rivals. Big oil firms are likely to go through a

shake-up that will make the merger waves of the past look like ripples on a millpond. The head of exploration for a supermajor thinks that by 2010, the industry will see another round of consolidation as stronger firms gobble up those unable to adapt to today's harsher environment. Other pundits are gloomier, arguing that even consolidation will not save the majors. Such folk argue that the once-proud giants may have to reconcile themselves to shriveling up over time as they fail to replenish reserves. They would then "hollow out" into technology companies not unlike today's Halliburton or Schlumberger and be mere handmaidens to the NOCs.

And yet, just as it is too early to declare the death of Detroit, it is probably too soon to count the majors out. The big boys of oil are extremely resilient, as their rebound from the nationalizations of the 1970s proved. They are still capital-rich, they command the top talent in the business, and they can still claim to have the edge in technology. Publicly, at least, the majors remain unbowed by today's challenges. With a self-confidence characteristic of senior Exxon officials, Henry Hubble dismisses the idea that the majors are handing back so much cash to shareholders because they have few ways to grow: "We saw the same arguments in the 1980s with high prices . . . for us, being selective is not restricting opportunities." BP's boss, Lord Browne, also insisted his firm has plenty of investment opportunities for the next decade.

Any survival strategy for the majors must center on technology for two reasons. First, that is an area where they already have strengths that the NOCs (bragging aside) do not. Second, the majors are likely to be banned from developing the cheapest and easiest reserves. The scraps left to them, such as squeezing extra oil out of marginal reservoirs and enhanced oil recovery techniques, are already proving to be the most technologically complex endeavors. Derek Butter of Wood Mackenzie, an energy consultancy, goes so far as to argue that "there are not enough opportunities for the largest majors in traditional oil exploration to replace reserves." That, he thinks, will force them to pursue such technology-intensive options as unconventional oil and natural gas.

With much of the conventional resource base closed to them, the majors are increasingly looking to unconventional hydrocarbons that they used to turn their noses up at. The best example is Canada's tar sands, mucky hydrocarbons that can be converted to useful gasoline at much greater expense, effort, and environmental damage than ordinary petroleum. In theory, there is more energy contained in Alberta's tar sands than in all the oil in Saudi Arabia. In practice, however, the stuff has proved so complex and costly to extract that Alberta's oil production today is still just a trickle compared to Saudi Arabia's.

Most of the majors are now plowing big money into tar sands, shale, coal-bed methane, and other such marginal projects, which Jeremy Leggett and other environmentalists fear will lead to a global-warming nightmare. Lee Raymond even famously boasted to have invested sufficient resources to "turn unconventional oil into conventional oil" in time. Given Exxon's track record, it may be unwise to bet against him. If governments want to tackle global warming in earnest, they should craft energy policies (like carbon taxes) that ensure carbon-intensive ways of making gasoline do not get a free ride.

Fighting Back

If the majors do face an unfair battle with subsidized rivals, one ironic possibility is that NOCs may eventually end up suffering as a result of the government patronage that today seems to be their unfair edge. A collapse in prices would deal them a devastating blow. David Victor of Stanford University argues that the NOCs "are usually so grossly inefficient" that their grand ambitions and hostility toward foreign investment are sure to fade when they are forced to survive without high export prices. Indeed, the majors could help that day of mutual cooperation arrive by softening their stance. Typically, they have insisted on majority ownership of assets and high returns on capital—"skimming the cream," in industry jargon. In the future, they may need to be more flexible. Christophe de Margerie, the boss of the French oil firm Total, likens this to a seduction: "You can no longer just say 'I am the king' and expect countries to give up their resources. You have to bring a win-win package, offering things like electricity generation, refining systems and training. You need to be sexy."

Ironically, a sharp rise in oil prices could also expose the folly of resource nationalism. Japan tried for many years to subsidize its way into oil and gas exploration; the bureaucratic misadventure proved so expensive and produced so little oil that Japan officially scrapped the effort early in 2005. Even CNOOC's Fu, who likes to see Chinese firms flourish at home, believes that the current infatuation with energy security overseas is misguided. He thinks the government backing so welcomed by NOCs today may eventually backfire: "Opportunities for oil come not where you pick them." China's friends may not offer the best prospects, he notes, and countries that are not friends may offer better prospects.

In short, oil is found not where governments wish it were but where on earth it happens to lie. And if oil really were to grow scarce in those few places like the Middle East in which it has historically been bountiful, then there would be reason for

alarm. That is precisely what a growing chorus, joining Campbell and the other early petro-pessimists, now argues: the problem of replacing reserves has grown so serious that this cannot
possibly be a problem limited to a handful of oil companies.
Rather, goes the emerging conventional wisdom, the troubles
at Shell and other publicly traded companies are a crystal-clear
sign that the entire earth is running out of oil altogether.

Of course petroleum is a nonrenewable resource, and so
by definition the Age of Oil must one day draw to a close.
But that day is much likelier to come as the result of carbon-
constraining policies and innovations in clean fuels and cars
that leave oil behind—not because the oil runs out. To see
why, consider the astonishing role played by technological innovation in the history of oil.

The Bottomless Well

"Oil is found in the minds of men." So says a bumper sticker
popular among petroleum engineers. True enough. In 1859,
Colonel Edwin Drake struck oil in Pennsylvania by drilling,
rather than digging, for oil; he adapted the old Chinese trick
of drilling for salt. That prompted the world's first oil boom,
which inevitably led to bust as oil flooded the market and prices
collapsed. In 1901, another set of unlikely innovators struck
oil in unpromising terrain near Spindletop, Texas. They did
so using novel drill bits that rotated through the earth, rather
than merely pounding it repeatedly, and so reached far greater
depths than were previously possible. That unleashed a fero-
cious gusher that spewed out hundreds of thousands of barrels
of oil in days and marked the birth of the modern oil industry.
Inevitably, this boom led to bust once again, as oil grew ever
more plentiful.

And yet, despite this history of innovation and abundance,

concerns about depletion once again are clouding the industry's future. It is not just a small handful of silver-haired gloomsters arguing the case for Peak Oil any longer; it is now close to conventional wisdom that this time around, the world really does face a looming crisis of depletion. And, the pessimists hasten to add, technology will not come to the rescue this time, as it has in the past.

Of course, this nonrenewable resource has to run out someday—but whether that peak has come and gone, as Deffeyes and Campbell maintain, or is decades off makes quite a difference. Official forecasters are fairly relaxed in their estimates of when global oil production will peak. The United States Geological Survey did a comprehensive survey of the matter in 2000 and concluded that such a peak is unlikely to come for two decades or more. The IEA takes a similar view, arguing that oil supplies will not be constrained until beyond 2030. In all likelihood, the world has a couple of decades in which to organize sensible public policies and stimulate investment in new energy technologies that can lead to a profitable and relatively painless transition away from petroleum. But if, on the other hand, the pessimists are proved right, it may already be too late to avoid disaster.

The pessimistic argument made by Campbell about the beer running out at the pub sounds reasonable, but it is wrong in two ways, one profound and one practical. The philosophical problem is that the pessimists fall into the trap of treating the level of recoverable oil resources as fixed—like the amount of beer in that mug. In fact, they are not fixed at all. Indeed, the expert estimates on the resource base for petroleum have consistently grown over the past few decades, despite the fact that the world has been guzzling oil during that time. We actually have more proven oil reserves today than we did twenty-five years ago.

The upward revisions are the result of the interplay of economics and innovation. The IEA describes this dynamic dance this way: "Reserves are constantly revised in line with new discoveries, changes in prices and technological advances. These revisions invariably add to the reserve base." A few decades ago, the average oil recovery rate from reservoirs was 20 percent. Thanks to astonishing advances in technology, that rate has risen to about 35 percent today. However, that still means that two-thirds of the oil known to exist in reservoirs is abandoned as uneconomical—leaving room for tomorrow's discoveries or innovations to lift recovery rates and magically push back the global peak even further toward the horizon. Some pundits had predicted that fields in the North Sea would peak in the 1980s. In fact, the peak was reached about fifteen years later.

Dozens of similar examples from around the world added up to defy Campbell's prediction of a global Hubbert's peak by 2004, a forecast that plainly did not come true. That may seem an unfair critique, as he had no way of knowing about the wave of offshore drilling technologies that would come about in the last decade. But that is the point—today's pundits cannot foresee tomorrow's innovations.

Petro-optimists say the future of oil is bright indeed. Peter Odell of Erasmus University in the Netherlands argues in a recent book, *Why Carbon Fuels Will Dominate the 21st Century's Global Energy Economy*, that these factors will ensure conventional oil will not peak until nearly midcentury and that unconventional oil resources (like Canada's carbon-intensive tar sands) will not peak until nearly 2100. Morris Adelman, the grand old man of oil forecasting and a professor emeritus at MIT, has even argued that the "amount of oil available to the market over the next 25 to 50 years is for all intents and purposes infinite."

A New Age of Discovery

There is another, more practical fallacy embedded in the gloomy forecasts too. "I challenge the idea that the era of discovery is over in oil," says Total's Christophe de Margerie. Thanks to the cold war and other political constraints on Western investment, much of the world has not been explored with the latest technologies. Russia is a good example. When Russia opened up to private investment under Boris Yeltsin, it suddenly saw an inflow of modern technology and management talent. The result was a breathtaking 50 percent surge in Russian production—a renaissance now put in serious jeopardy by Vladimir Putin's crackdown on the sector. Nick Butler, formerly of BP, even sees eastern Siberia as the great "new frontier" for the oil industry.

Similarly, other parts of the world are still underrigged and underexamined. Says Fu Chengyu, president of China's CNOOC oil company, "Our offshore prospects are just beginning. A promising area the size of two North Seas has yet to be explored." India opened up its oil exploration sector to foreign companies, and Britain's Cairn did not waste much time before striking oil in Rajasthan. V. K. Sibal, India's director-general for hydrocarbons, gushed, "I expect much more, maybe even a 'supergiant' deep offshore somewhere near the waters off Myanmar."

The unexplored potential in the Middle East is vast. Pete Stark of IHS Energy Group, a leading consultancy on this issue, says that Iraq has over 130 undrilled prospects; his firm released a study in 2007 suggesting Iraq may have double the oil reserves currently on the books. Neighboring Saudi Arabia has about 260 billion barrels of proven oil reserves today. Its oil minister, Ali Naimi, is confident that current and future technologies will help him lift that figure by 100 billion barrels over the next few decades. He points in particular to an unexplored region on the Saudi-Iraqi border the size of California.

Total's de Margerie points to frontiers opened up by technology, not just politics: "OK, there may not be any more glamorous Ghawar fields, at least onshore, but there is tremendous opportunity if we look at 'deep horizons.'" He points out that there are large deposits 30,000 feet or more underground. The snag is that they are usually under very high pressure or temperature and may have an extremely acidic composition. As technology improves, he thinks extracting "these very strange hydrocarbons" will become economically viable: "In the past, nobody looked there because the technology did not exist."

Similarly, the industry is now pushing into water depths that were unimaginable a decade or two ago. In the Gulf of Mexico and elsewhere, oil rigs now float atop 10,000 feet of water. These marvels of engineering are stuffed with the latest in robotics, electronic sensors, and satellite equipment. Their spotless control rooms are a far cry from the grimy gushers of Spindletop. Using fancy multilateral wells that twist and turn every which way, they are able to hit giant underwater pockets of oil miles away from the rig. So there are lots of frontiers—to explore and to exploit—left. However, just because there is plenty of oil left on earth does not mean it will get to market easily and cheaply. On the contrary, the global oil industry faces a daunting task. It will require extraordinary innovations, and the courage and capital to back them, to bring the world's remaining hydrocarbons to market.

The Spirit of Spindletop

Hence the petro-pessimists' second great doubt: that Big Oil has run out of techno-fixes. They make three arguments. First, technological advances like multilateral wells are proving a double-edged sword, leading to faster decline rates at reservoirs. Second, they argue that there are no more "killer applications" like 3-D seismic surveys left to transform the exploration industry. Finally, they argue that the majors have largely abandoned the vital task of investing in upstream research and development in recent years as part of their misguided cost-cutting drives, and so no longer have the capacity to innovate. This is a more serious critique than the one involving Hubbert's peak, as it cuts to the heart of what will make or break the oil majors. De Margerie challenges the two strands of petro-pessimism: "The peak will come, but we can keep the plateau for a long time with technology."

Consider the notion that technology could be a double-edged sword. It is true that in some fields, such as complex offshore projects, the majors have recently been surprised when investments in fancy technologies have led to spurts of increased output ending in faster depletion. Therefore, argue some, these technologies merely act as fatter straws, helping suck out more liquid now but ultimately emptying the glass more quickly. Roger Anderson of Columbia University has looked for this alleged "faster depletion effect" in over forty oil and gas fields using the latest innovations and has found no evidence of it. He argues that "the more prevalent problem is not that there is faster depletion, it is that oil companies desperate to get the 'black gold' into the bank are ignoring modern asset management techniques." He points to the example of firms using advanced "4-D" seismic production technologies but failing to tie the production of oil and gas to the market and price conditions in real time. That is as absurd as "Toyota making sure it had 'just-in-time' assembly lines but ignoring consumer purchasing and pricing signals."

Besides, a few examples should not be generalized across all technologies and fields. That argument falls into the same trap as the argument of those who say recoverable resources are fixed; a fatter straw could end up producing both more oil now and more later if the resource base is seen as dynamic. In most cases, modern techniques clearly prolong a field's life and increase the ultimately recoverable resources.

Andrew Gould, chairman of Schlumberger, a French-American oil services giant, points out that twenty-five years ago, only one-sixth of exploration wells were successful, while two-thirds are today. Over that time, the success rate for development wells has risen from about 33 percent to nearly 100 percent. He is convinced that the future lies in embedding digital technologies such as down-hole sensors, real-time communications equipment, and other fancy tools that add up to the smart oil field of the future. Companies already use some of these techniques today when they drill wells, but he argues the key is to use them for monitoring wells "from day one." "Progressive illumination" was the management philosophy of the past: "You learned as you went along. Now, you draw a much better picture up front, and monitor the reservoir carefully from day one." Private-sector companies do not want to spend such money up front, at least not yet, but he points approvingly to Saudi Aramco's long-term thinking.

Deep in the windswept deserts of eastern Saudi Arabia, there is a petroleum visualization center on par with the best in Houston. Backing it up is a bank of computers with more data storage capacity than America's NASA. Unlike most private companies, Saudi Aramco has invested in observation wells that monitor its reservoirs in real time. Company engineers say with some satisfaction that they can monitor what is happening in a well deep underground from a laptop computer anywhere in the world. Using this fancy technology, the company's geologists say they are able to ward off the problems of field decline that Matt Simmons has made so much noise about.

The Central Bank of Oil

"Aramco is a peculiar company," says a smiling Abdallah Jum'ah, the chief executive of Saudi Arabia's state-run oil firm, as he greets a visitor at its stylish, modern offices in the country's oil capital, Dhahran. Inside the heavily guarded compound is a tidy community of houses, schools, and baseball fields that would do suburban America proud. Foreigners and locals mix freely, and the canteens serve Budweiser (albeit the nonalcoholic kind). Women even drive cars, in defiance of Saudi custom, though they must get in the backseat when they reach the edges of the sprawling Aramco gated community and let a chauffeur take over.

Suddenly this most secretive of firms is opening up in other ways, too. Having been widely blamed for this decade's soaring oil prices, the embattled giant hopes to prove it can deliver on its most important promises—and survive the terrorists who seem bent on destroying it. Aramco is not publicly listed and avoids debt, so it has no need to divulge much to the financial markets. The best outside guess in 2006 was that it produced eleven million barrels of oil per day—over one-eighth of the world's consumption—with annual revenues of well over $100 billion. Not only does Aramco sit atop the world's largest reserves by far, it also enjoys the world's lowest discovery and development costs—about $0.50 per barrel, one-tenth or less of what private-sector rivals pay in Russia, the North Sea, or the Gulf of Mexico. With at least 260 billion barrels of proven oil reserves left, Aramco is twenty times the size of ExxonMobil, the largest private-sector oil firm.

Forecasters assume that Aramco will have to double its output in the coming decades to meet the expected growth in global demand. Yet some vocal outsiders doubt that it can do so. Led by Matt Simmons, a Texas investment banker, they argue that Saudi oil fields may already be facing technical difficulties. Some 90 percent of Saudi oil comes from just seven fields, whose average age is forty-five to fifty years. Just one, Ghawar, produces a staggering five million barrels per day, at least. Simmons claims that academic studies, as well as troubles at fields in neighboring countries, suggest that Aramco's giant fields may soon go into decline.

As a result of such criticisms, Aramco's technical experts now openly discuss field data previously held secret. Its geologists explain exactly how they intend to maintain an output of fifteen million barrels per day or higher for fifty years—even without the new oil discoveries that they insist could eventually add another 200 billion barrels of oil reserves. Far from suffering from a greater need for water injection to maintain reservoir pressure, Ghawar appears to be stable for now, thanks to Aramco's sophisticated use of technology. Nansen Saleri, Aramco's head of reservoir management, says, "We released more data in 2004 than we did in the previous fifty years. On a field-by-field basis, we now release more than the investor-owned companies." The exception, of course, is Royal Dutch/Shell, which had all its field data audited by outsiders in the wake of a major reserves overbooking scandal. Of course, Aramco would go a long way toward debunking the doubts raised by Matt Simmons if it were to follow Shell's lead on outside auditors.

Jum'ah deserves credit for opening up the firm, and the evidence he has produced is reassuring, but when asked if he will allow an independent audit of that evidence, a glint of steel appears in the Saudi's eyes: "Why should we? We have never failed to deliver a single barrel of oil promised to anyone, anywhere." That is a big boast, but a defensible one. The Saudis have, over the last two decades, clearly acted as a "central bank" of oil, releasing oil from its buffer of spare production capacity onto the market during times of war or crisis. During the Iran-Iraq War, the first Gulf war, the Venezuelan national crisis in 2003, and the second Gulf war, the world market suddenly lost one to two million barrels per day of oil output. The result would surely have been a sharp spike in oil prices, perhaps even panic or an economic shock—if not for the swift and silent action by Aramco to release a tidal wave of oil onto the market. Jum'ah invoked that history to prove that Saudi Arabia is a reliable and responsible partner in the oil economy—but that same evidence also highlights exactly how much the world relies on the stability and good graces of the desert kingdom's ruling elite.

Only time will tell whether Matt Simmons is right in his claims about Saudi reserves, but he is clearly right to demand full transparency and outside scrutiny of Aramco. Given that the world economy's future hinges on the reliability of Saudi oil production, simply saying "Trust me" won't cut it for much longer. The lack of transparency at Aramco neither proves nor disproves the depletion hypothesis, but it does suggest something else. The risk of Saudi implosion—due to geological inadequacy or political instability—is surely as good a reason as any to wean the world off oil.

Innovate or Die

What of the critique that there are no more breakthrough technologies left to transform the oil business? After all, the spread of seismic visualization technologies has completely transformed the oil business. The daunting task of identifying and characterizing promising oil reservoirs, work that used to take a team of engineers and geologists weeks or even months, is now done in a matter of minutes or hours. There is now more rocket science used in the oil industry than at NASA, thanks to all the supercomputers, advanced software, visualization technology, and the like. By one estimate, the industry saves roughly $11 billion a year thanks to its use of 3-D seismic survey technology.

The industry is divided on this matter. Some in the business say there is no obvious blockbuster technology on the horizon. They point to interesting but lesser ideas. For example, Exxon and Schlumberger are investigating whether adding electromagnetic analysis to seismic soundings will improve the visualization of reservoirs. Apache is investing in technology that allows 3-D visualization without the need for big amphitheaters or special goggles.

Peter Robertson, vice chairman of Chevron, says, "I would not bet the company on a new 3-D seismic." He is convinced that incremental technologies matter because they can help lift recovery rates a few percentage points and improve recovery in existing fields: "Flattening the decline curve could mean more than even a big new discovery." Halliburton's boss, David Lesar, has not given up hope, though. He argues that "when 3-D seismic or directional drilling first came, nobody saw their potential. It was the unexpected and unanticipated application of those technologies that was key." He thinks today's innocuous technologies could prove tomorrow's breakthroughs, as long as the industry continues to encourage innovation.

That points to the most explosive criticism leveled by the petro-pessimists at the oil majors: that they no longer have the capacity to innovate. A few decades ago, of course, these firms were fiercely proud of their proprietary technologies, which they believed gave them a competitive edge. But during the 1990s, most majors slashed funding in this area, leaving service firms like Schlumberger and Halliburton to pick up the slack. This is not so different from what happened with the big Detroit car firms, which have outsourced much of the development process to specialist design firms and the suppliers of parts and even entire subassemblies. In oil, this "hollowing out" of the majors has clearly opened the door to nimbler independent firms and outsiders like NOCs—and that same trend is lowering the notoriously high barriers to entry for the automobile industry too.

"Ten-dollar oil killed upstream research," says one executive. Ivo Bozon of McKinsey estimates that the majors slashed upstream R & D spending from $3 billion in 1990 to below $2 billion in 2000. Over that same period, the service companies increased research investments from $1.1 billion to $1.7 billion. Bozon adds that the sharpest cuts were among American companies. Small wonder there is much resentment among petroleum engineers and geologists in Houston's oil patch. "These guys need to explore, but they don't know how to do it anymore," complains Roice Nelson of Geokinetics, which makes reservoir visualization software for the oil industry. Nelson helped found Landmark Graphics, an industry pioneer in imaging software, so his criticism stings. He notes that the industry sacked many of its most qualified technical staff and that relatively few college students are going into petroleum engineering or related fields. "We'll be working till we're past eighty," he sighs.

The majors are now realizing that this shift away from technology, once their core strength, was a mistake that has benefited three sorts of rivals: the service companies, the "mini-majors," and the NOCs. Halliburton's David Lesar is delighted by it: "There's been a fundamental shift in ownership and development of technology from the majors to the service companies. Unlike in the past, this is the place to innovate!" The problem is that the service companies are less capable of investing for the long term, since their balance sheets tend to be weaker than those of the majors. This shift is not unlike the move in the auto industry to outsource technology and innovation to the "Tier 1" suppliers, which tend to be financially weaker even than the troubled Detroit automakers.

The shift in innovation has been a boon to smaller oil companies, which are not so risk-averse. Especially since the wave of mergers, the majors need megaprojects with long lives to replace reserves. That has made them extremely wary of trying new technologies. Says Chevron's Robertson, "Taking a flier on a project with a very long lead time and very high investment" is simply too risky. Contrast that with the attitude of Steve Farris, Apache's boss: "We go to the service companies and say 'What have you got?' Hell, we'll spend money to try it." The spirit of Spindletop lives on, it seems.

All this hurts Big Oil in another way, because the NOCs no longer need them to get access to modern technology. The more sophisticated NOCs, like Saudi Aramco, buy technology directly from the service companies, but many others are turning to the "indies" for help. Jim Hackett, chief executive of America's Anadarko, explains that with a market capitalization of $20 billion and a capital budget of $3 billion a year, his firm is large enough to challenge the big boys: "I can't compete with Exxon in twenty countries, but I can beat them in a few." Aside from the speed of decision making and the embrace of technology, he thinks resource nationalism gives the smaller Western oil firms an advantage these days. "We are no threat, we have no baggage of the Seven Sisters. Sometimes locals don't even know that we are an American firm."

Whether the majors will regain their skills as technology innovators is an open question. Exxon, for one, is making a big push. The firm spends some $600 million a year on upstream R & D, more than its rivals. And it sees technology as the key to unlocking future reserves. Of course, merely investing in new technologies for extracting ever more complex hydrocarbons is no solution. The sorts of innovations needed to tackle the new geopolitical and environmental realities of the energy world are genuine alternatives to oil.

There are some positive signs. Other majors are now following Exxon's lead in investing in alternative fuels, striving to turn mucky tar sands and shale oil in Canada and Venezuela, as well as cleaner alternatives like ethanol from sugarcane and corn, into useful fuel that can be blended with conventional refined petroleum to "make" a form of gasoline. This is a breathtaking shift for an industry that has doggedly defended petroleum as a feedstock—and undermined many an effort to boost alternative fuels. As this shift happens, the great oil exploration companies of yore may well transform themselves into *manufacturers* of motor fuel. That could be good if it led to a big breakthrough in the development of clean alternatives to oil—but, as Big Oil is no doubt aware, it could just as easily help stretch out the world's remaining reserves of petroleum for many more decades and make it all the harder to kick the oil habit. And as Jeremy Leggett warned by the German bonfire, unless public policies push the industry to embrace clean fuels rather than dirty ones, the dam could break away from biofuels and toward carbon-intensive tar sands and the like. But either way, this trend toward the manufacture of fuel further debunks the Peak Oil argument, as it makes it even less likely that the world's gas guzzlers are going to run out of fuel anytime soon.

Concentration, not Scarcity

The upshot of all this is that while there may be a short-term scarcity of oil or gasoline on the market from time to time in coming years and the depletion arguments are worth keeping an eye on, there is no compelling reason to think the world is actually going to run out of fuel for your car tomorrow. The real trouble with oil arises from carbon and *concentration,* not scarcity. There is good reason to think there is plenty of petroleum left in the world—it's just not in the hands of the investor-owned oil companies with household names like Chevron. The rise of NOCs makes it much harder for the Western oil majors to replace reserves and to serve their useful role as an anti-OPEC. But that's just the beginning of the geopolitical problems arising from oil.

The growing concentration of oil reserves in the hands of a few countries creates several big problems for the world. As allegations of "blood for oil" surrounding America's invasion of Iraq suggest, the foreign policy mischief that has long been associated with petroleum will only get worse in coming years. High oil prices and resource nationalism are also strengthening the hand of autocratic rulers from Russia to Venezuela, much to the detriment of democracy and economic progress in countries with oil. Most alarming, the inexorable rise in market power of a few autocratic and erratic petro-regimes in the Persian Gulf strengthens OPEC's grip on the world economy and sets the stage for the next great energy shock.

One big difference from the two big oil shocks of the 1970s is that America, the biggest of all oil consumers, is at last starting to prepare the ground in advance for what is to come. The next section of the book examines the question of whether the dinosaurs of the oil and car industries can really learn to dance. If there is to be a renaissance for these industries, it will come only because foreign firms lead the way. American oil companies in particular have their heads in the sand on climate change. That shortsightedness may yet come back to haunt them if the automobile industry perfects technologies for clean cars that allow it to leave its troublesome twin behind. The next chapter chronicles the rise of Toyota, which is not only leading the world auto industry, knocking GM off its number-one perch, but also showing the way for the auto industry to break free of its troublesome dependence on oil.

II

CAN DINOSAURS DANCE?

While America's oil and car lobbies rely on political clout in Washington to defend their status quo, Toyota and BP are proving that even dinosaurs can learn to dance.

CHAPTER FOUR

The Parable of the Prius

How Toyota's culture propelled the once-provincial carmaker past GM to number one

The Detroit Auto Show hit the nation's front pages back in 1992, when the president of Chrysler Corporation entered the convention hall by driving a Jeep Grand Cherokee through a plate-glass window to draw attention to its revived product line. American car firms often pull stunts during this annual expense-account celebration, which is a highlight of the car industry's year. Toyota, in contrast, tends to be a bit less flashy. That is because its cars speak for themselves.

For two weeks in January, the town of Detroit can convince itself it is still Motor City, capital of the car world. The North American International Auto Show, as it is formally called, attracts approximately seven thousand journalists, a third of them from over sixty countries overseas, making it one of the biggest media events in America. It is also a reflection of how the car industry has become global. The visitors to the Cobo convention center in downtown Detroit file past the hunched statue of the city's local hero, the scrappy fighter Joe Louis, in the foyer. They gaze at some seventy new models that the world's car manufacturers choose to unveil at the event. By the time the show closes, about 750,000 members of the public will have visited it; a further 7 million will have tuned in on network television.

The show has been going since 1907, when a new century saw the birth of a new industry, one that was to define both an era and a country, America. Now the industry's future is being driven by companies from Asia and Europe, while the American contenders are in retreat. Toyota has steadily overtaken General Motors as the world's leading automaker. The twin hegemony of oil and autos is also under attack. Concerns about global warming from the carbon-dioxide emissions of internal-combustion engines, plus worries about the security of energy supplies from petroleum reserves in the volatile Middle East, have ensured that. The race is on to find ways around these problems. Some of the contenders will be the stars of those industries today. Excitingly, however, the podium could feature newcomers capable of setting the pace and changing everything.

The same old auto show stunts keep getting repeated. In January 2006, Tom LaSorda, boss of Chrysler, drove another Jeep through another specially prepared glass wall before crossing the road to have it scale an artificial minimountain in front of the old Detroit firehouse. Later that evening, he and his German bosses from DaimlerChrysler donned waiters' aprons to pull jugs of draft beer and dish out hamburgers in the firehouse. The firehouse bar opened its doors five months after September 11, and while the food and drink flow, $10 bills donated for firefighters' charities rain into special glass urns. When a bigger bill hits an urn, someone rings the old station bell. Toyota eschews all this showbiz: it is a different kind of carmaker. Toyota is where the lessons are to be learned for the future of the industry, for the future of personal transportation, and for dealing with the problems of pollution, energy supply, and global warming that go with the unbridled use of autos.

Contrast the typical Detroit hoopla with a different scene in the River Ballroom at the Cobo Center. It is a Monday, the second day of the 2006 show. The room is packed for the unveiling of the "all-new" sixth generation of the Toyota Camry, the best-selling car in America since 1997. Conceived for the American market, it was first introduced in 1982. In the twenty-four years since then, it has sold some 10 million models, over 6.5 million of them in America. The latest version, said the sales chief presenting it, was a radical departure, pushing the limits of the "design envelope."

The wraps are pulled off the three cars onstage to reveal . . . a nice new Camry, with styling a little sharper than its predecessor, but still the comfortable, tidy, well-equipped Camry that has driven its way into the hearts and garages of suburban America. The audience look at each other, shrug their shoulders, and mock the extravagant words of the warm-up: "It's no design revolution; it's another Camry," says one Detroit veteran who looks as though he has been there since the beginning of the Camry story—perhaps even since the 1960s, when Detroit's Big Three made nine out of ten cars sold in America. The Camry Mark VI looks solid and well made. It drives well, and it does all it says on the can. You can select an engine to suit your needs. That is why American families love it. No great fuss about the number of horses under the hood or the sizzling acceleration. Yet it has all the things Detroit has forgotten how to do with cars: comfort, style, reliability, and reasonable looks. That is how Toyota became the world's largest car company and the number three even in Detroit's home market. Welcome to the

new world of the American auto industry—a world where globalization laps at the very feet of the Big Three. For half a century, they expanded and tried to extend their hegemony around
the globe; now the world has invaded their home territory and
taken control. One of them has had to be rescued painfully by a
European group. By the middle of the first decade of the 2000s,
it became an open question whether the remaining two could
survive independently, without foreign assistance.

While GM and Ford have for years teetered on the edge of
bankruptcy, with quarter after quarter of losses, Toyota's profits topped $10 billion in 2006. On the stock market, it became
worth more than GM, Ford, and DaimlerChrysler put together;
it was worth even more than the combined value of its successful
compatriots, Nissan and Honda. Only the Korean manufacturer
Hyundai joined the Japanese Big Three in earning a real return
on capital—the French pair Renault and PSA Peugeot Citroën
occasionally does that as well, but rarely sustains that level
of performance. The success of Toyota shows up everywhere.
Whether it's the Camry consistently topping the sales and quality charts, or Cameron Diaz and Tom Cruise turning up at the
Oscars ceremony in its little Prius gasoline-electric hybrid cars,
or its opening of new American car factories faster than Detroit
can shut old ones, Toyota has been making waves. It has been
surging past GM to become the world's largest manufacturer of
cars. After seventy years as the industry leader, GM has been
overtaken by a Japanese company whose products appeared in
America only in 1958, to widespread derision. With America
gripped by mounting panic over its dependence on imported oil
from the politically combustible Middle East and many states
growing anxious about global warming, Toyota is doing more
than any American car company to address these concerns.

Toyota has taken over the past, commands the present, and

is leading the way to a brighter future in which cars are powered by unconventional means. It is fathering the technology
that will save the industry and America from slavish dependence on petroleum and reliance on resources from unstable
parts of the world.

In the mid-1990s, it looked as though Daimler-Benz was
the first car company to latch onto the importance of cutting
carbon emissions, with its pioneering work on fuel-cell–electric
cars. But Daimler's boss, Jurgen Schrempp, and his chief technical adviser, Ferdinand Panik, headed off down a blind alley
with one narrow version of hydrogen-fuel-cell cars based on the
cumbersome conversion of methanol, a dangerous, dead-end
fuel. That mistake, combined with the strain on the company of
assimilating Chrysler after 1998, doomed that pioneering effort
to expensive failure. The Toyota approach was quieter, more
careful, and ultimately more successful.

The hybrid technology in the Prius and in other, grander Toyota products, such as the luxury Lexus, is one of the treatments
administered to wean America off its dangerous addiction to
oil. By reducing fuel consumption and developing the hybrid
technology that will one day help drive zero-emission hydrogen-
fuel-cell cars, Toyota is demonstrating technological prowess to
match its newfound industrial leadership. By mid-2006, Toyota
had sold nearly double the number of Prius hybrids—around
500,000—that Ford Motor Company could sell across its entire
range of hybrid models, which exists only thanks to licensing
the Japanese technology. Detroit has scrambled to latch onto
Toyota's success in exploiting a whole new segment of the car
market for Americans who care about the environment and
want to make their own contribution to dealing with global
warming. Toyota did this because it has the long-term vision
to address environmental issues ahead of the pack, but also

because it could: it had the industrial, financial, and technological strength, vastly increased in the past ten years, to undertake the task. That is the parable of these two cars: the Camry gives customers what they want today, while the Prius promises to deliver what they will need tomorrow.

A decade ago, all this looked unlikely. The past ten years have seen the decline of Detroit and the triumph of Toyota—the biggest and richest of Japan's Big Three, ahead of its compatriots, Nissan and Honda. As recently as 1995, it was all so different: then GM was still top dog, challenged by Ford, while Toyota was trailing behind, in third place. Shortly after becoming Toyota's president in 1995, Hiroshi Okuda, a swashbuckling samurai-like figure among the gray-suited ranks of Japanese executives and the first nonfamily member to take the helm at Toyota, was asked which other car companies he admired. He said, "On the technical side it is Mercedes-Benz. On finance it is GM because of its skillful management of cash flow and assets." This very un-Japanese frankness made sense at the time. GM had pulled back well from the brink of bankruptcy, in 1992. Three years later, its profits per car were much greater than Toyota's, even though the Japanese company's factories were more efficient. In 1995, GM made a trading profit of $9.8 billion, compared with Toyota's $3.3 billion; GM's profit margin on every car was 5.8 percent of the selling price, while Toyota's was 3.2 percent. But beneath the surface were undercurrents that would soon sweep away American leadership. However, few could see them, far less understand what was really going on.

The Big Three all grew weaker because they did not pay enough attention to new technology, to developing innovative new cars, and to maintaining the quality of their output. They thought that a profusion of brands and extensive dealer networks could keep them on top. They had mastered the SUV market, and they thought this would ensure their dominance. They had not the slightest clue as to what was really happening to them.

Even as they slid down the rankings for product quality and consumer satisfaction, the companies tended to blame external factors for their ills. Competition from Japanese producers was somehow unfair, they insisted. The yen was greatly, deliberately undervalued, they squawked. And that was even before the blame started to be pinned on Detroit's labor unions. It was before the argument emerged about American carmakers being saddled with high labor costs, including pension and health-care benefits—as if somehow these deals had not been signed by Detroit bosses themselves in happier times, when making money seemed easy. Instead, it was all the fault of an unfair world, in the eyes of Detroit.

Blame a Cruel World

Start with the beleaguered boss of GM. A graduate of Harvard Business School and Duke University, where he was a basketball star, Rick Wagoner started in the car company's finance department in New York City. Wagoner got onto the board as finance chief and right-hand man of Jack Smith, who was made chairman and chief executive of the company in a boardroom coup in 1992, after the world's biggest manufacturing company had suffered a close brush with bankruptcy. At one point, top executives were clustered around a fax machine, waiting for the latest verdict on the company from a debt-ratings agency. Another downgrade would have pushed up the company's cost of borrowing and forced it to seek court protection from its creditors under Chapter 11 of the bankruptcy law. Wagoner was a key man in the rapid recovery that made GM's figures look good again, to the point where it was briefly a benchmark for

Toyota. Despite his sterling financial training and his contribution to that earlier GM recovery, Wagoner comes up with a highly skewed account of recent history to explain how Detroit's finest fell back into financial disarray.

As chairman of GM, he never gets to his feet without including in his speech a reference to the fact that the Japanese government intervenes in the world's currency exchanges to depress the value of the yen and help the exports of big Japanese manufacturers, such as Toyota. Governments do from time to time interfere, using their money to adjust currencies. The most significant example was the Plaza Accord in February 1985, when finance leaders of the rich world sat down in a New York City hotel to work out a way to lower the value of the dollar against the major currencies in Europe and the Japanese yen. The idea was to bail out America, which was then running a huge trade deficit, by having other currencies go up in value against the dollar by selling dollar reserves they held in a coordinated swoop on the exchanges. A cheaper dollar would in turn make American goods relatively cheaper in world markets, thereby boosting sales and plugging the trade gap.

The ploy worked; the dollar fell by half in the following years, and America's exports came close to balancing with imports for a while. While the Plaza deal kept the world economy moving along and corrected the imbalances in America's trade deficit, it set the scene for an economic squeeze in Japan, one that was to galvanize Toyota into international action to compensate for the weakness of demand at home. Selling the dollar to bring down its value had the side effect of boosting the money supply in Japan. This meant there was plenty of money around, and interest rates were low. This in turn led to a huge stock market and property bubble in the late 1980s. Land prices in downtown Tokyo skyrocketed. The 5 square miles of land around the emperor's palace were soon worth more than all the land in California, which is physically bigger than the whole of the Japanese archipelago.

In 1989, land prices collapsed, along with shares on the Tokyo Stock Exchange. Some $20 trillion of asset value was wiped out in the space of a few months. Japan's economy started to decline, and prices fell, even after interest rates were cut to the bone. Car sales fell from six million a year to barely four million after soaring during the bubble. Companies such as Toyota were up against the wall. This was the time when Nissan's descent into crisis began in earnest. Toyota, led at first by Eiji Toyoda, one of the firm's founding family, and then by Okuda, decided to look abroad for salvation. Since exporting cars from Japan was becoming more difficult as the yen's value rose in the wake of the Plaza Accord, the firm's leaders decided to build more factories overseas, in Southeast Asia, America, and Europe. It would take time to build up this overseas production to the point where today, Toyota makes more cars abroad than it does in Japan.

In the mid-1990s, most of the cars Toyota sold in America were shipped from Japan. Okuda became the first nonfamily leader of Toyota, in 1995. He is a big, broad, muscular man, standing out from most of his countrymen, although his athleticism could not be more Japanese: he has a black belt in judo and even into his sixties looked like he could literally floor any antagonist. After the *Economist* ran a story about a renegade Toyota sales manager, one Konan Suzuki, spilling the beans about the company's methods, Okuda greeted the author with a hearty backslap and a reprise of the offending article's headline: "Konan the Barbarian, ho-ho."

This larger-than-life character immediately set the company the task of becoming competitive even as the Japanese currency rose in value to the point where it took only 95 yen to buy a dollar; previously the exchange rate was around 150 yen to the dollar. This was the currency reality that affected Toyota and, indirectly, GM. It was quite the opposite of Wagoner's view of history. Squeezed by a stagnant economy and an insufferably strong exchange rate, Toyota raised its game and made itself more efficient simply to survive—by competing more successfully in world markets. That is the clue to the real truth behind the rise of Toyota and the fall of GM and the rest of Detroit: Toyota had learned how to make very efficiently cars that appeal to consumers, who keep coming back for more. Instead of having a top-heavy corporate structure with lots of divisions and different brands and dealer networks, like Ford or GM, Toyota kept itself quite simple. Its basic models such as the Corolla and the Camry sell around the world, with only minor modifications for different markets. That way, it gets huge economies of scale. Toyota did not set out to become the biggest car company in the world. It just aimed to survive—but in doing so, it became the best.

Foreigners often make the mistake of thinking all Japanese companies are the same, behind a mask of a foreign language and an opaque culture. But this is not true. Whereas Nissan, before its reinvention under Carlos Ghosn and Renault since 1999, was very much a Tokyo company, its head office full of suave graduates of Tokyo University, Toyota was born in Aichi Prefecture, 250 miles southwest of the capital. Its bosses have always retained a slightly provincial air of plain speaking and rough manners. For a long time, huddled in a cluster of factories in what is called Toyota City, about thirty minutes' drive from the bustling commercial hub of Nagoya, it was the most inward-looking and Japanese of the country's car companies. Even today, its head office is there, with only a token office building in Tokyo. Until quite recently, many of its factory workers in Aichi were not even the typical Japanese "salaryman." Instead, they were part-time farmers, growing rice or oranges in their smallholdings between shifts at the six big car plants that gradually gobbled up more and more of the land. In the 1970s, these Japanese assembly-line workers found themselves on two sides of the growing Japanese-American trade war. By day, they worked for Toyota, fighting to get its imports into America. By night, they were trying to hold off cheap American rice and oranges by supporting Japan's agricultural tariff barriers against imports. No wonder Tokyo sophisticates used to call Toyota a "bunch of farmers."

This homespun culture for some time held back Toyota, which relied solely on exporting vehicles made in Toyota City and in other plants built in Japan's underdeveloped southern islands, where labor was cheaper. This was a different situation from that faced by its two main competitors. Honda had quietly spread itself around the world, following the market progress of its motorcycles, which always served to awaken consumers to the brand before it tried selling them cars—just as it had begun life as a motorbike maker in Japan before expanding into four-wheeled vehicles. Nissan had decided early on that it needed factories in Europe and America to serve those markets while getting around import tariffs and other protectionist barriers, such as import quotas.

Nissan was a slick but bureaucratic business, comfortable with its own entrenched traditions. Honda was more like many American firms—always a little bit the outsider, dominated by clever engineers who rose to the top by designing flashy motorbikes. Bikes led the motorization of Japan, as of so many other countries, after World War II, and Honda rode that boom to greatness, transferring its engineering know-how into making cars notable for their great little engines and sporty handling.

Beside them, Toyota was always the dullard, its products boring and its image provincial and dowdy. Its strength in its home market came from its superior manufacturing (which Nissan and Honda soon copied, from the 1970s onward) and from a mighty marketing and distribution network. Car buyers did not go to Toyota. Toyota went to them. Long before Detroit and its dealers had ever learned about customer relationship management (CRM), Toyota had been practicing it. Its representatives for years have literally been knocking on doors and inquiring about a family's automobile needs.

In the early days, files of index cards held all this data, so that the visiting salesman would know that a child was graduating from university and would want his or her first car, even before he knocked on the front door. The sales rep from Toyota was the Japanese equivalent of the Avon lady or the man from Prudential. The index cards have been replaced by computers, but every Toyota dealer still has mountains of data on customers—what they bought when, and so on. The company to this day retains several different distribution chains that it controls, each one aimed at particular groups of consumers.

Okuda was soon to change Toyota dramatically. At a personal level, he quickly became famous for saying exactly what he thought, a staggering break from Japanese corporate culture, where there is really no word for "no," and no euphemism or evasion ever goes unuttered. His two big worries were making the company competitive even with a strong yen and avoiding trade disputes with Europe and America. He feared that the company's growing success in dealing with the former would serve only to bring on the latter. He worried about a resurgence of protectionism in America and Europe. Okuda called it his fear of "trade friction." He had seen what rows over Japanese imports had done in the 1980s, leading to import restrictions, so-called voluntary restraint deals that the Japanese had to tolerate even in the markets of such free traders as Ronald Reagan and Margaret Thatcher.

To deal with this, Okuda came up with a plan to convert Toyota from an export business based in the Japanese backwoods into a full-fledged global multinational. So he decided to double the size of production at Toyota's factory in Derby, England, and to build a new assembly plant in Valenciennes, in northern France. The French move was also taking a leaf out of Henry Ford's book. Ford, right back in the 1920s, opened factories outside America on the grounds that you had to make cars where you sold them. Until Toyota started making cars in their country, French people scarcely knew what a Toyota was, much less bought one.

Next in line was America, where Toyota had shared a factory in California with GM since the mid-1980s. Toyota had already opened other plants, notably in Kentucky, but Okuda knew that if "trade friction" were to be avoided a second time around, the company would have to drive further. Thus it opened a factory in Princeton, Indiana, in 1996; another in Huntsville, Alabama, in 2001; and a third in San Antonio, Texas, in late 2006. Okuda's sensitivity to protectionist worries also led to Toyota opening a technical development center in Ann Arbor, in the green hinterland of greater Detroit. The other reason for this was that, despite the declining fortunes of the American Big Three, Detroit remains a world-class center of automotive technology, with a large pool of trained engineers and technologists. It remains a cluster, a sort of Rust Belt Silicon Valley, even as the number of manufacturing jobs declines. The same is of course happening to Silicon Valley, where chip manufacture is now overshadowed by bigger volumes in Asia.

Toyota: Made in the USA

The key question in this globalization of Toyota, Japan's leading car manufacturer, is How come it is so much better than GM or Ford, and how come it can export this expertise all around the world. The answer lies in Toyota's management approach, which emphasizes a learning culture, an eagerness to embrace change, and a relentless pursuit of excellence. Rather than merely deal with problems, the company's employees are hardwired to find out the underlying causes and to tackle those.

After all, Henry Ford and Ford Motor Company invented modern mass manufacturing, all the way from the use of standardized parts to the moving assembly line. Thirty years ago, Americans and Europeans, enduring a rising tide of imported Japanese cars, thought the reason was that cowed Japanese workers slaved away for long hours before crawling home to snatch some rest. One senior European Commission trade official in the 1970s, after a visit to Japan, famously described Japanese car workers as toiling like hamsters on a wheel and sleeping like rabbits in cheap hutches. Such nonsense was dismissed only when the Japanese first opened factories in America, notably the Honda plant in Marysville, Ohio, and the joint Toyota-GM plant in Fremont, California. The big lesson from these transplant factories was that American workers could perform just as well as Japanese ones, in work conditions that were perfectly acceptable. So what explains the difference between Japanese success and American failure?

Start with Toyota's history and the world into which it emerged. It may seem surprising, but Toyota's success owes a lot to America. It all started when an American naval officer, Commodore Matthew Perry, in 1853 sailed into Tokyo Bay with his so-called black ships to force the Japanese authorities to open up trade. This affront undermined the power of the ruling Tokugawa shogun ruler, because it demonstrated the weakness of the country under his rule. The ensuing unrest led to the Meiji Restoration fifteen years later, in 1868. This was really a revolution that reinstated the emperor in an effort to make the country stronger and to protect Japan from the "foreign demons." The capital was moved from Kyoto to Edo, a port city on the east coast, which was renamed Tokyo. Japan started opening up. Under the control of a handful of oligarchs, with the emperor as a figurehead, Japan began importing Western technology in a bid to modernize and build its military power, leading later to wars with Russia and China.

Toyota was founded as a textile machinery business during this period. The car company, Toyota Motor, was not formed until 1938, and made only a handful of cars, on a craft basis, before it was assigned to build various military vehicles during World War II. It restarted carmaking after the war. By 1950, Toyota had turned out only 2,685 cars over thirteen years, at a time when Ford's main Rouge plant in Dearborn outside Detroit was making seven thousand a day. Under the occupation rule of America's General Douglas MacArthur, the economy was kept tightly controlled by credit restrictions to limit incipient inflation. This led to a slump in demand for cars and the need to lay off 1,600 workers in 1949.

The American authorities had brought in labor laws that gave plenty of power to trade unions as part of the effort to promote democratic institutions after the militaristic imperialism of the previous thirty years. Toyota faced a disastrous strike and factory sit-in that brought the young car company to its knees, as it ran out of bank loans. The solution worked out between the owners, the Toyoda family, and the unions, was for the company president Kiichiro Toyoda to resign, taking responsibility in a typically Japanese manner for the misfortunes that had befallen the company. A quarter of the workforce was fired. Those who remained were promised lifetime employment, but in return they had to agree to do any work that needed doing and to be thoroughly flexible. Thus, the death knell was sounded for Ford's mind-numbing assembly line, as Toyota gave birth to what is now known as flexible manufacturing.

One huge example of this was the way Toyota started using production workers to perform maintenance and other tasks, such as realigning the production line for changes in product. There is one striking early example that shows the genius of Tai-ichi Ohno, Toyoda's personal manufacturing guru, for spotting simple solutions to difficult production problems. Way back in the early 1950s, he made a huge breakthrough when he found a way to change the giant 300-ton presses in the body shops that press out the panels that are welded to form the floor, walls, and roof of the basic monocoque car body. With plenty of money for investment, the big manufacturers could afford to have separate press machines for different parts. This meant fewer changes of dies by specialist contractors that usually took a whole day to perform, during which time production workers were standing around idle—wasting time and money.

Ohno simplified the process by using rollers rather than cranes to maneuver the huge dies. This simplified the whole process to the point where it no longer required specialized contractors, and the production-line workers could be used to perform the work. Before long, they could do in three minutes what previously had taken a whole day. Just as important was the demonstration of the principle of flexible working that was a benign legacy of Toyota's traumatic strike in 1950. Needless to say, the company has never had a strike since, and flexible working is one of the hallmarks of Toyota assembly lines.

MacArthur had by this time done something else to shape the economy of Japan that was to prove even more determinant in the rise of Toyota. He had smashed the formidable prewar industrial trusts known as *zaibatsu* that had carved up the Japanese economy among them. Some of these old names—Mitsubishi, Mitsui, Sumitomo—survive in a watered-down form known as *keiretsu*, but the smashing of overbearing *zaibatsu* holding companies meant the way was open for young, dynamic companies with names such as Sony, Panasonic, Honda, and Toyota.

Beyond Henry Ford

The nascent Toyota was keen to learn from the rest of the world. Kiichiro Toyoda had visited Henry Ford's Rouge plant, then the most efficient in the world, back in 1929, and his nephew Eiji was to follow in his footsteps in 1950, when the company was casting around for ways to become more efficient. But the doctrine of Fordism—hard, repetitive, boring, but well-paid work, with high labor turnover and hirings and firings to fit the rise and fall of demand—could not work in Japan. One basic reason was that Toyota (the family name Toyoda meant "abundant rice field" in Japanese, and they had it changed to Toyota; the new name meant nothing then, though now it could stand for a relentless quest for excellence) now had labor as a fixed cost. So ways must be found to make best use of it, and hire today, fire tomorrow Fordism was not an option. So Eiji and his manufacturing expert Taiichi Ohno started developing a different way of working, taking advantage of the flexibility inherited from the deal that settled the strike. The contrast with

Detroit was impressive from those days onward. The strength of the United Auto Workers was also to make labor a fixed cost in America. But the Big Three continued with their antagonistic approach, with tough bargaining every few years for a new pay deal, with one company picked on for a showdown, leading to a new contract that would then set the pattern for the other two companies. There was no attempt to align the interests of labor and capital.

The Japanese approach at Toyota could not have been more different. Workers were known as associates, and even the humblest among them was to be treated with respect. The union would share the company goals of surviving and prospering. Unlike similar labor relations introduced by the occupying Allies in German factories, such as Volkswagen, provision was made in Toyota for the fluctuation of wages in line with the performance of the company. The British-designed system in Germany did cement social peace and help create the country's postwar economic miracle, but it was to prove ultimately inflexible. Toyota's foresight was to use the American-inspired labor regime to build in flexibility. A large part of the workforce pay would depend on results, giving the company a cushion to absorb shocks from the market and protect it from the cost of a fixed pool of labor. The pay of individuals would still be based on seniority, a tradition in Japan, giving the company an incentive to develop the skills of workers as they grew older in order to extract better value from them. The line workers would accept responsibility for quality instead of having supervisors check their work.

In this respect, Toyota was able to draw on the work of another American, W. Edwards Deming. An academic and statistician at New York University, Deming had been drafted by the U.S. government to perform various tasks during World War

II. Assigned at one point to the task of helping turn America's industrial might into a powerful war machine, turning out vehicles and the famous Liberty ships (cheaply made cargo vessels churned out at a rate of one a day to replace the thousands sunk by German submarines), Deming got interested in the use of statistics to ensure quality. Anybody could see that the cheapest way to make things in mass production involved getting each job right the first time, obviating the need for the disruptive and expensive repair of defects found at the end of the process. Deming's genius was to employ statistics as a tool to make that happen.

In 1950, as part of the occupation effort of putting a democratic Japan back on its economic feet, Deming was invited to teach in Tokyo. There he preached his gospel of using statistics for quality control by recording quality performance and analyzing it, then altering processes to improve quality and measuring the change and its outcome in turn. And on it would go, with repeated refinement, measurement, alteration, measurement, and so on, in pursuit of total quality. At Toyota, this gave birth to the doctrine of *kaizen*, or continuous improvement. To this, Ohno added the concept of elimination of waste (*muda* in Japanese). This meant waste of time, materials, energy, and even money by having brought-in components lying around for a long time before they were needed. Again this was an American import. Like Eiji Toyoda, Ohno visited Ford factories in Dearborn in the late 1940s. The founder of Toyota, Kiichiro Toyoda, had always been a great admirer of Henry Ford and made all his managers read Ford's book, *My Life and Work*. Indeed, in the late 1930s, the two companies had contemplated some sort of joint venture, until Pearl Harbor cut short their discussions.

But Ohno was surprised to find how old-fashioned the Dearborn plants seemed. There seemed to be nothing but huge machines and great piles of inventory around them, with everybody busy feeding the machines to keep them working at a great rate so as to reduce the cost of each piece stamped or cut or ground. But after each operation, the parts lay around for days, until great forklift trucks moved them to another pile before they were processed again. Ohno concluded that the young upstart Toyota could do better: David would show Goliath a thing or two. In those days, Toyota was making only about forty trucks a day, while Ford's Rouge plant alone was making eight hundred. Despite the volumes churned out by the most famous factory in the world, by the firm that invented the mass-production car industry, Ohno was unimpressed.

Yet something else did catch his interest. Ohno had never seen a supermarket before he went to America. Instead of the small mom-and-pop corner stores, to this day still prevalent in most Japanese towns and suburbs, he saw these vast emporia, with their shelves of products being continuously stacked. He noted the nonstop flow of deliveries and the way goods moved quickly from delivery truck to store shelves, with little time stacked in backrooms. When stocks of loaves or milk on the shelves went low, new supplies were wheeled in. Ohno had a sort of epiphany: he saw in a flash the possibilities of applying the same principles to manufacturing. He realized that the same principle could be applied to Toyota car factories. Translated from retailing to manufacturing, this became the basis of just-in-time production.

So there evolved the Toyota Production System (TPS), based on three principles: dedication to quality, to the elimination of all sorts of waste, and to continuous improvement. The linchpin of the war on waste was the system of *kanban* for the supply of parts, based on the lessons Ohno learned walking the aisles of

American supermarkets. The aim is to supply parts only as they are needed, rather than have them piled up in stores, forming an expensive inventory. The way it works on the shop floor is that when one bin of parts is emptied, another has to be arriving to take its place. This is what made the expression just-in-time or lean manufacturing, shorthand for TPS. Its architect was Ohno, inspired by the techniques of Deming and by the example of Michigan supermarkets.

The idea is that the demand for parts pulls production through the system, rather than building up interim stocks at every stage. It was brought to a new level in the 1970s, when Ohno hired the young Fujio Cho to be his assistant in disseminating the message through Toyota's whole operations. That means applying lean techniques to logistics and administration by endlessly trying to simplify and refine business processes, doing only what needs to be done, and then performing that function as efficiently as possible. On the production line, small teams of workers operate under a team leader, and tasks are rotated frequently to limit boredom. Teams meet after every shift in rooms at the side of the assembly lines to talk through problems and to come up with solutions. Another key aspect of the system is that quality is the responsibility of every member of every team. So, when a worker sees something going awry, he pulls a cord, known as an *andon* cord, that brings the line to a halt. This draws from Deming's analysis: if a fault is identified and rectified on the spot, if necessary by senior workers descending like a flying squad to fix things, the problem will be contained before the factory has churned out thousands of defective cars or car parts.

Walk into a Toyota factory in Kentucky, Derby in Britain, or Valenciennes in France, and you will recognize the same features—the same visual displays telling everybody what state production is in, what rate cars are being produced at. There will be the same jingles as the *andon* cord is pulled or to signify a change from one model to another or to signal that a stage of production has been completed, and the cars are moving on to the next station. Everything is synchronized; the work follows the same cadence from one end of the line to the other, with finished vehicles rolling off the line at the rate of about one a minute. No one rushes or appears hassled. Everything is spotless; there are numerous swings and hydraulic devices to take the heavy lifting out of moving subassemblies onto the developing vehicle as it moves along.

Cho was also charged with applying the techniques throughout the company's burgeoning international empire. When Toyota opened its huge plant in Georgetown, Kentucky, Cho went to run it. Although he barely spoke English to begin with, he would walk the lines daily and listen to workers. In 1980, Toyota had eleven factories in nine countries; in 1990 it had twenty in fourteen countries; by middecade, it had forty-six plants in twenty-six countries. In addition, it has design centers in California and on the French Côte d'Azur, plus engineering bases in Michigan, Belgium, and Thailand. Back in Tokyo, Cho and Okuda looked at the instant miracle performed at Nissan after Renault bought a stake in 1999 and parachuted in Carlos Ghosn, formerly Renault's chief operating officer, and a score of the French company's brightest young managers. Ghosn, within three months of arriving in July 1999, laid out a plan to turn around the company in little more than a year. Within nine months, it was clearly beginning to work. He closed five plants, got rid of 25,000 workers, sped up product development, hired a new design chief, and raised cash to pay off Nissan's debts by selling stakes in the parts makers Nissan owned through its involvement in the *keiretsu* system. Ghosn, a charismatic

Brazilian-born Frenchman, likes laying out clear objectives and moving fast. His way of working is to appoint cross-functional teams drawn from different departments and get them to come up with solutions to problems. "I believe creativity comes at the interstices between different functions," he says.

Toyota bosses were so impressed that they launched their own offensive to squeeze out yet more savings, particularly from the suppliers of parts. Toyota has long had an efficient system of assigning the supplier the task of supplying a part to perform to certain standards, then leaving it to the company to work out the design, the materials, and so on. Toyota cares only that the part or subassembly does what it is supposed to do and meets its quality standards by not breaking down. This is critical to the company's success, since up to 70 percent of the cost of a Toyota car can be in the parts. Under Okuda and Cho, the company worked to simplify manufacture, even taking a leaf out of Chrysler's book: Chrysler, in the early 1990s, brought out a Neon model that was famously economical to make by having, for instance, doors made up of fewer panels and parts than normal. Back in Toyota City, engineers quickly tore apart an early model to see what they could learn from America.

It was to be a long time before knowledge flew in the opposite direction. The astonishing thing about the collaboration between Toyota and GM in the Fremont, California, plant was how little the Americans learned from it. Individuals did, but they were not able to transfer their knowledge into the mainstream of GM. But a group of business academics led by the Massachusetts Institute of Technology (backed by most leading car companies) was carrying out a five-year, $5 million program of research to benchmark car manufacturing operations around the world. Out of that emerged a book written by James Womack, Daniel Jones, and Daniel Roos. Called *The Machine*

That Changed the World, it was an instant success, because it described Toyota's manufacturing system in some detail and set it in the context of an analysis of what was going on elsewhere in the world car industry. This was the first really independent look at TPS. For the first time, carmakers elsewhere could compare themselves with Toyota and set up their own lean manufacturing processes to begin the catch-up process.

The problem for the competition was quite simple: they were chasing a moving target. Okuda and Cho knew that even they were not fully exploiting their capabilities. They decided to make a determined bid to be leaders in the world, deepening their penetration of the world's biggest car market, the United States, and extending their operations throughout Europe and Southeast Asia. Of the three-million-car increase in output of the global car industry since 2000, about half has come from Toyota as it pursued its aim of Global Ten, winning 10 percent of the world market. Then Okuda set his sights on 15 percent, a target that would take the company to the level of GM. It was well on its way to hitting that target.

A Toyota Way

But there is more to Toyota's triumphant rise than just a world-class manufacturing system. After all, Honda was ahead of its bigger rival in building up a global network of fully flexible factories capable of making different models according to demand. And carmakers in Europe and America have long since adopted Toyota-style lean practices. Yet nobody does it quite as well as the original masters. There is a good reason for that, beyond the fact that Toyota was the first mover in raising the standards of efficiency and quality in modern car production. Its continued edge in manufacturing is a reflection of a deeper cultural feature of the company.

There really is a "Toyota Way." This set of guiding principles was only formally put into print in recent years by Cho. Jeffrey Liker, professor of industrial engineering at the University of Michigan and a veteran student of Toyota, identifies no fewer than fourteen guiding principles. Written down, they sound commonplace, even trite. They range from basing decisions on a long-term approach, regardless of short-term costs, to an emphasis on getting processes right and then continuously improving them. They include the use of lean "pull systems" to avoid overproduction and the buildup of wasteful stocks, and working steadily, more like a tortoise than a frantic hare. Most important, the Toyota Way stresses developing the skills of people to add value to the organization. Again, this latter trait goes back to the lessons learned in 1950. Since workers were promised lifetime employment, it was in the company's interest to develop their skills so that they contributed more as they got older, rather than just relying on their muscle power. It is a sort of social contract between management and workers. But none of these fine aspirations would mean anything if they were not instilled in Toyota's people and in their ways of working with each other.

Tetsuo Agata, in the later years of his long career, doubled as general manager of Toyota's huge Honsha plant in the Toyota City complex and as the company's overall manufacturing guru. Before that, he lived for years in America and has many American friends. One was a senior manager with Dow Chemical, and they often talked about their respective employers. For Agata, Toyota was best summed up for him by his friend, who observed that Toyota people such as Agata always "kept outside the comfort zone." Whenever they hit one target, they set another. In other words, the world-beating TPS is but the most visible aspect of a company culture of relentlessly embracing change, tackling challenges, and pursuing excellence.

For Jim Press, head of Toyota in North America, the company's secret is all about the focus on the customer. As a young man growing up in Kansas, he had always assumed he would become a car dealer like his uncle, who sold Chevrolets. He and his best friend, his wife's sister's husband, both went to work for Ford in California on the customer-relations side, Press thinking it would give him good experience for becoming a dealer.

Besides, California in those days of surfing and Beach Boys fever seemed an attractive alternative to the plains of Kansas. When his Ford boss left to join Toyota, which was just then setting up its American marketing operation in Torrance, on the outskirts of Los Angeles, in 1968, Press followed. He had already become a little disillusioned with the way Ford handled customers. He saw a chance to do things better in the fledgling Japanese importer. He recalls Henry Ford II saying at that time that Ford, with cars like its little Pinto, would pretty soon drive the Japanese back into the sea. Ford's bombast soon collapsed when the Pinto began bursting into flames in even minor rear-end collisions and had to be withdrawn. For Press, "It was all a great adventure. I didn't think it would last, but at least it would be a good experience." Last it did. Press can still remember the excitement at Toyota in 1970, when their sales passed 100,000 cars. At that time, the only big import brand was Volkswagen, which enjoyed sales of around 400,000 a year. VW was to lose its early lead by never really establishing a strong manufacturing base in the United States.

On the stand at the Detroit Auto Show in January 2006, the day before the new Camry launch, Press was the man unveiling the Lexus 460, a version of the luxury sedan that rocked America when it was first introduced at the show back in 1989. That was the year when the Detroit Auto Show became the North American International Auto Show, and the launch of the Lexus and Nissan's rival Infiniti underlined the internationalization of what had been a largely insular industry.

Lexus was a brand conceived for the American market and has only recently been launched in Japan. Nothing could better illustrate the Americanization of Japan's leading car company. Toyota now sells more cars in America than anywhere else, and the American market contributes about two-thirds of company profits. The Lexus was the idea of Toyota managers at its American headquarters in Torrance. Although the company's Camrys and Corollas were runaway hits, it irked the bosses that more affluent consumers drove Mercedes or BMW models.

Top management back in Toyota City was soon persuaded to launch a product that could compete head-on with the best Mercedes cars. Naturally, it had to compare favorably with the German luxury cars in terms of performance, ride, handling, and style. One detail shows the thoroughness with which Toyota engineers pursued this goal. Determined to have an engine that was smoother and quieter than those of competitors, they went to the heart of the problem: what auto engineers call noise, vibration, and harshness (NVH) emanating from the engine. Normally, luxury cars dampen all this with insulation under the hood. Toyota went one better and designed an engine with parts engineered to such precise tolerances that there was less NVH to start with. This is a perfect example of the Toyota management approach: do not just deal with a problem, find out the underlying cause, and fix that. If the assembly line stops, find out why; if the answer is because a part is late, find out why; if it is because the delivery truck failed to show up in time, find out why; and so on until the root cause has been identified and action has been taken to prevent any repetition.

The development of the Lexus car and the whole Lexus brand was a radical step, especially for such a cautious company as Toyota. Its arrival was a loud wake-up call to Detroit: it showed there was nothing the Japanese manufacturers could not tackle. To their humdrum, reliable, everyday sedans was added the very best in affordable luxury. The boss of Britain's Rolls-Royce, David Plastow, recalls returning to London flabbergasted at the quality Toyota was offering for the price. In effect, he had glimpsed the future that was to undermine the independent existence of the most famous luxury carmaker in the world. Rolls-Royce and its sister brand Bentley have long since been carved up and sold off to BMW and Volkswagen.

Here's to You, Mrs. Jones

As he strode the platform extolling the virtues of the new model, Press was like a smooth preacher bringing the room to God. Discussing the secrets of Toyota's success, he adds a quieter, but equally evangelical tone to his remarks. "The Toyota culture is inside all of us. Toyota is a customer's company," he says. "Mrs. Jones is our customer; she is my boss. Everything is done to make Mrs. Jones's life better. We all work for Mrs. Jones." Press says he always liked the idea of delivering customer satisfaction. "It was nice to go to parties, tell people what you did and have them say that they liked those cars."

Press picked up on the inherent Japanese interest in conservation and the environment back in the early 1970s. As a corporate officer with senior responsibilities in Japan as well as America, it was no surprise to Press to see hybrids moving up the agenda. The Prius began life as the result of an uncomfortable feeling that Toyota's chairman Eiji Toyoda had in the late 1980s. The Japanese economy was booming, sales were strong, the Lexus had been launched to great acclaim in America—the market that was coming to matter more than any other to the company. He feared Toyota was growing complacent and would be caught out when the bubble burst. So he tried to instill a sense of crisis in the company. He asked the top research and development engineers to consider the sorts of cars that might be more suited to the twenty-first century than traditional models.

The blue-sky thinkers kept coming back to growing concerns about natural resources and the environment. So a group was formed to develop Global 21, the working title for what became the Prius. At first there was no question of its being a hybrid. The initial requirement was simply for a vehicle that could offer 50 percent better fuel economy than the basic Corolla model, which did 30 miles to the gallon. A senior test engineer, Takeshi Uchiyamada, was put in charge, even though he had never worked in vehicle design. This was a deliberate move, because Toyoda wanted the company to overhaul its product-development process, and G21 was to be in more than the literal sense the vehicle for that process. It was to lead to Toyota's *obeya* system: *obeya* means "big room" in Japanese. This new way of working was to bring all the relevant people together in a sort of control room, rather than have the chief engineer and his designers go around to different technical experts separately. The engineers were then asked in late 1994 to speed things up and develop the G21 as a concept car for the 1995 Tokyo Motor Show. The bar was also raised: now the target for fuel economy was to be a 100 percent improvement on the Corolla.

Agata too remembers when Okuda and Cho decided to increase the pace of development of more fuel-efficient cars in the mid-1990s. Engineers had been working away quietly on Toyota's long-term plan to develop hydrogen-fuel-cell cars. These combine hydrogen gas with oxygen from the air to produce water and electricity, which then drives the car. There is a lot of complicated electronics involved in converting and controlling the electric current and applying it to the wheels. Even though they are fuel cell–powered vehicles, they are essentially hydrogen-electric hybrids. So, recalls Agata, when the boss decided that Toyota should come up with a car that had between one and a half and twice the fuel economy of conventional gasoline engines, engineers turned to the fuel-cell program and extracted the hybrid technology. "We felt that with the time horizon so far away for practical, economical fuel cells, it would be sensible to bring out an interim product," says one vice president of Toyota Europe.

Now the Toyota Prius has its critics. Many consumer tests show that it was better at delivering fuel economy in government tests than in ordinary motoring. The French manufacturer of Peugeot and Citroën cars in particular claims that in everyday driving, its fuel economy is little better than one of its diesels. Meanwhile, Peugeot has been quietly working on a diesel-electric hybrid to combine the benefits of that fuel's economy with the electric boost from a hybrid. GM and Ford at first derided the Prius's achievements before switching to bringing in hybrid versions of their own gas-guzzling SUVs, licensing Japanese technology.

As well as bringing public-relations kudos for being seen as doing something concrete about the environment, the Prius is teaching Toyota a lot about how to develop and package fuel-cell power trains when the core fuel cell has been brought down to an economical production cost and made robust enough for everyday use. But that alone was not the reason for introducing the Prius. Toyoda, Okuda, and Cho foresaw mounting pressure for carmakers to be seen to be doing something about the need for greater fuel economy as a way of addressing concerns about global warming. The hybrid formula was simply Toyota's best shot at achieving their objective: Cho and Okuda decided to press ahead. As a result, Toyota has taken the lead in the race to develop the successor to the internal-combustion engine, the car of the future. Thus the machine that changed the world (Toyota Production System) is producing the car to do it again: the first mass-produced, alternative car to conventional, internal-combustion engines—the Prius hybrid.

The essence of the Toyota story is that it is a company with a learning culture. It is a world-class heuristic device, a machine for promoting continuous self-education of the individuals and of the company culture. It first learned how to make mass production work more harmoniously than anywhere else in the world, moving on from the successes and limitations of earlier pioneers such as Henry Ford. It then learned how to turn itself into a multinational company, with a tight focus on succeeding in the biggest and most profitable car market in the world, North America. It did so largely by learning to listen to the heartbeat of American consumers. The secret of Toyota's American success lies much in little things. One example is the teams of designers sitting for days on end in supermarket parking lots, studying the way Americans loaded and unloaded their cars. Such assiduous attention is as important as grand strategy.

The Prius as Parable

Now Toyota has learned that the automotive world has to move on from the era of the gasoline internal-combustion engine. The hybrid Prius is not the car of the future. It is the machine that will teach Toyota and others about what will be the automotive future in a world challenged by carbon emissions and oil's geopolitical complications. Toyota is not alone in spotting that something drastic will have to change over the next twenty years, as the world gropes toward carbon-constrained, oil-free (and quite possibly hydrogen-powered) motor transport. Detroit has ideas, too, but its business situation is so precarious that it lacks the money to keep up the pace—and even seems shortsighted enough to derail efforts aimed at helping it along.

In the early 1990s, Al Gore pushed a federal program called the PNGV—Partnership for a New Generation of Vehicles. Detroit took the money, but nothing much emerged apart from one-shot vehicles such as an all-aluminum Ford Taurus that never went beyond prototype. Critics say PNGV was just a cynical sham—the federal government pretending to talk about doing something big about pollution, fuel economy, and carbon emissions, and Detroit pretending to listen, happy to collect the money and go along with the exercise. Toyota and the Japanese were excluded. Ironically, that put the fear of God into them and helped steel the determination of Toyota's bosses to make their own, more serious effort. In this, as in so much else, they have left Detroit trailing in their wake.

The value of Toyota's contribution is that it will drag the rest of the world's auto industry along with it. When the industry leader decides that it is time to look for a world beyond petroleum, then the others have to follow suit. For the first few years after the Prius appeared, Detroit bosses mocked it as nothing but a publicity stunt before they eventually saw the point and scrambled to join the race to fuel the car of the future. Their initial reaction was a grave misjudgment of the Japanese and the Toyota way. Toyota decided around the time of the first Gulf war that there was a long-term trend toward concern about the environment and about the availability of resources. It decided to start doing something about it. Okuda and Cho, the men running the company in those days, looked at the changing map of world oil. They could see the tilt of reserves back to the Middle East.

Japan is a country that went to war once, bombing Pearl Harbor to safeguard the oil supplies needed for its imperial expansion in the Pacific in the 1930s and '40s. As the postwar Japanese economy emerged as a rich consumer of resources such as oil, the first oil shock rocked that country more than any other. The Japanese thought their postwar miracle was being snatched away by greedy oil producers in OPEC. With no indigenous resources and no energy of its own (apart from some uneconomical coal), Japan was once again vulnerable. Amid rising panic about oil imports, the government made great efforts to promote conservation and to switch production from heavy industries such as steel and shipbuilding into electronics. This was the time Japan went from being the shipyard of the world to become the cradle of the consumer electronics industry, nudged along by farsighted government policies that recognized painful long-term trends in a way that America's energy policies have yet to recognize global warming.

So it should be no surprise that Japan and Toyota should be in the forefront of reacting to the new, worrying world of international energy. The troubling geopolitics of oil is at long last pushing other governments to look for alternatives to petroleum. Big Oil bosses are reviled and ridiculed, and their companies seen as dinosaurs. Saudi Arabia is widely seen today as a fount of economic instability, terrorist financing, and political uncertainty. Some even claim it is an unreliable oil supplier and untrustworthy American ally.

But things were not always so. Half a century ago, universal access to petroleum was seen as the essential guarantor of political stability and Middle Eastern oil viewed as the great hope for the world's economic development. The Seven Sisters were seen as courageous firms usefully advancing America's foreign policy goals overseas. As the next chapter chronicles, America embraced the young country of Saudi Arabia and its vast petroleum bounty back then, thus forging one of the most powerful, persistent, and problematic geopolitical alliances of all time: the great Axis of Oil.

CHAPTER FIVE

The Axis of Oil

Oil's geopolitical complications arise from America's bipartisan addiction to oil

Toward the end of World War II, Franklin Delano Roosevelt attended a summit that changed the course of world history. No, not the famous meeting at Yalta, with Joseph Stalin and Winston Churchill. Immediately after that gathering, Roosevelt traveled quietly to the USS *Quincy*, anchored in the Red Sea. The man he met there had rarely set foot outside his home country and insisted on bringing an entourage of dozens, which included his household slaves and a royal astrologer. He did not preside over a vast nation, he was not a significant world ruler, but he was in command of the greatest store of the one commodity the modern world needed more than anything else: oil. He was King Ibn Sa'ūd, father of the new country of Saudi Arabia, home to the biggest reserves of oil on the planet, then as now.

That this meeting should take place in the closing months of World War II was not surprising. Unlike World War I, which was a tragic story of static conflict, blood, and mud, World War II was one of movement. Trucks, tanks, and all sorts of

vehicles played a central role in most battles. Access to petroleum was woven into every strand of strategy. The Japanese
bombed the American fleet at Pearl Harbor in part to try to
break the embargo that was limiting its supplies of oil and rubber from Southeast Asia. Hitler's mad, fateful dash to Stalin-
grad was partly to secure access to Caspian and Romanian oil.
Rommel was driven out of Africa by the British and American
armies, lamenting the fact that lack of gasoline had limited his
maneuvers. Oil had emerged as a major strategic element.

In the years before the war, the desert kingdom had gone
from sleepy backwater to the most promising oil province in
the world. U.S. military planners were painfully aware of the
swift decline in domestic oil reserves—for much of the early
twentieth century, after all, America was the Saudi Arabia of
the global oil world. The impending peak of American production alarmed the national security experts, who knew all too
well that the country's military and economy had been built on
the assumption of plentiful and cheap petroleum.

They scoured the world for a possible replacement for American reserves, and they struck upon the new and untapped bounty
in Arabia. The energy and defense gurus briefed FDR about
the paramount importance of that oil. So the man who courageously defended global democracy and led America to victory
over fascists in Europe and Asia made a Faustian bargain. In
return for guaranteed access to Saudi Arabia's vast quantities
of oil, Roosevelt promised the tribal chieftain America's full
military support. Revealingly, the oil-for-security guarantee was
extended not to the people or the government of Saudi Arabia
but to the Al Sa'ūd ruling clan itself.

In the decades since, that vow and the resultant alliance between America and Saudi Arabia has proved to be one of the
few fixed points of global politics, more durable even than the

cold war. Call it the Axis of Oil. This perversion of foreign policy in oil-consuming nations is one of three powerful ways in which oil affects geopolitics. Oil also corrodes the democratic and economic development of countries endowed with it, often pushing them toward autocracy or state failure—a phenomenon known as the Oil Curse. The final link between oil and global politics is one that haunts the world economy: the prospect of an oil-induced energy shock.

Don't Mention the "O" Word

Oil muddles geopolitics to the point that it has become a serious national security threat. Of course, the destabilizing geopolitical effects of petroleum were not always so apparent. When FDR struck his compact with King Ibn Sa'ūd, he was genuinely convinced that the free flow of Saudi oil would be good for the Saudis and good for the world economy. And to be fair, America was not the first great power to intermingle petroleum and power politics. A young Winston Churchill, on the eve of World War I, made the risky decision to convert the British naval fleet from Welsh coal to imported oil. The resultant gains in speed and maneuverability gave Britain a decisive edge over Germany in that war—but also set off a global scramble for oil supplies that led Britain to colonize what is now Iran (and to set up the Anglo-Persian Oil Company, the precursor to British Petroleum, now called simply BP).

Even so, the times have changed, and it is now blindingly obvious that reliance on petroleum brings with it undesirable and unnecessary geopolitical tangles—and yet, the Washington policy machine marches on as though the Axis of Oil still rules supreme. With the cold war long over, there is no longer any plausible ideological justification for America propping up thuggish dictators in the oil-rich part of the world. And things will get only worse in the next decade or two, as non-OPEC reserves dwindle. Serious problems are in store. Two-thirds of the world's proven reserves of conventional oil lie in the hands of five countries in the Persian Gulf, with Saudi Arabia atop one-quarter of the world's reserves. As oil gets depleted rapidly in other parts of the world, the West *will* come to depend ever more upon these currently undemocratic and perhaps unreliable countries.

The Bush administration's invasion of Iraq has led many to suspect that it's all just "blood for oil" for the two oilmen in the White House. The problem with that Michael Moore-esque view of the Iraq debacle is that it ignores an even more profound conspiracy at work: how the country's addiction to oil has unduly influenced American foreign policy, regardless of the party controlling the White House. FDR was a Democrat, of course, and so too was Jimmy Carter—the first president to declare publicly that American foreign policy in the Persian Gulf was to defend the free flow of oil "by any means necessary." The militarization of America's energy policy has been a bipartisan affair.

Giving in to the facile demonization of the Bush energy policy also lets Congress off the hook too easily. Regardless of which party is in charge, the energy lobby has found a reliable friend in the Washington establishment. For many politicians of both parties, it is a reason to call for energy independence. But the phrase has become misleading, for it is used to justify subsidies for pork-barrel projects or mere sops to the industry, such as drilling for oil in the Alaskan wilderness. Given that America consumes a quarter of the world's daily production of oil (and nearly half its daily production of gasoline) but has barely 3 percent of its proven reserves, it will never be energy independent until the day it stops using oil altogether.

The Paradox of Plenty

An unusual meeting took place in October 2005 at Saint Matthew's Catholic Church in Baltimore. After the sermon, some parishioners stayed behind to hear two African emissaries on a national tour sponsored by Catholic Relief Services, a religious charity group concerned with poverty issues. The two speakers took turns explaining the harm that America's gasoline-guzzling does to the poor in faraway lands. They spoke eloquently about foreign policy abuses; of wars fought and blood spilled; of corrupt dictators kept in office; and especially of the helpless, impoverished masses kept in poverty, which they argued was the result of America's misguided energy policy.

The audience listened politely. When the speakers finished, one elderly African American parishioner raised his hand and let the visitors know in the gentlest way possible that he wasn't buying the guilt trip. He said very sweetly, "Most of us Americans know about Africa only through what we see on television. I know Africa is very rich in diamonds, gold, and oil, but the people are very poor. So why are your governments so bad at managing that wealth?" Austin Onuoha, a human rights activist from Nigeria who just moments before had been railing against American profligacy, smiled and conceded, "You hit the nail right on the head."

Economists have long observed that developing countries that are rich in oil tend to do surprisingly poorly when it comes to sustaining economic growth. In a landmark study in 1995, Jeffrey Sachs (a prominent development economist at Columbia University better known these days for his work on poverty) and his colleagues showed that the resource rich grow more slowly than other poor countries—even after such variables as initial per capita income and trade policies are taken into account. That was a stunning conclusion, for it flew in the face of intuition and apparent common sense: how on earth could a country with no natural resources grow faster than one endowed with plenty—especially black gold?

The usual explanation for this is "Dutch disease," named for the hardships that befell the Netherlands after it found North Sea gas. When a country strikes hydrocarbons, a sudden inflow of dollar-denominated revenues often leads to a sharp appreciation in the domestic currency. That tends to make nonoil sectors like agriculture and manufacturing less competitive on world markets, thus leaving oil to dominate the economy. Experts have offered fixes for the economic aspects of this Oil Curse for a while. Some governments have used stabilization policies: when oil prices are high, revenues are set aside; when prices fall, governments use the funds to cushion the blow. A related idea is to park part of the proceeds from resources in offshore "funds for the future." In theory, such funds would not only help spread the wealth over several generations but also help avoid overappreciation of the local currency. Some countries even disburse some oil revenues directly to every household. Alaska and the Canadian province of Alberta use some variation on this idea of direct disbursements, which has the advantage of ensuring that ordinary citizens, and not just oil bosses or crooked politicians, benefit directly from the resource wealth that belongs to everyone.

These are fine ideas in principle, and in developed countries they even work, to some extent. But charges of wasteful government spending and cronyism abound. It has been suggested that a new Iraqi government could disburse oil bounty to its citizens, too, but doing so properly in a country where many do not even have bank accounts will be tricky. Norway has an offshore oil fund that is often touted as a model for developing countries. Yet even the virtuous Norwegians have occasionally raided it for politically popular causes. In developing countries, such raids are the rule rather than the exception. Zambia set up a stabilization scheme to manage mineral exports; but as prices soared in the 1970s, the government dropped it—and years of pain followed when prices fell again. Venezuela set up a fund for the future, "El Fondo de Inversiones," in 1974, but was soon raiding the kitty. An orgy of domestic spending left the country with a herd of white-elephant projects, huge foreign debt, and declining social spending.

Michael Ross, a political scientist at the University of California at Los Angeles, argues that oil-rich countries do far less to help the poor than do countries without resources. He points to evidence that oil- and mineral-rich states fare worse on child mortality and nutrition, have lower literacy and school-enrollment rates, and do relatively worse on measures like the UN's Human Development Index. Economics offers some explanation of why this is so. Unlike agriculture, the oil sector employs few unskilled people. The inherent volatility of commodity prices hurts the poor the most, as they are least able to hedge their risks. And because the resource is concentrated, the resulting wealth passes through only a few hands—and so is more susceptible to being pilfered by rulers and well-connected elites.

No Taxation Without Representation

This misdirection points to another explanation for the Oil Curse that is gaining favor: politics. Because oil money often flows directly from Big Oil to the Big Man, as Africa's dictators are known, governments have little need to raise revenues through taxes. Arvind Subramanian of the International Monetary Fund (IMF) argues that such rulers have no incentive to develop nonoil sources of wealth, and the ruled (but untaxed) consequently have little incentive or ability to hold their rulers accountable.

Some thinkers, especially indignant officials from Persian Gulf countries, argue that their region has escaped the Oil Curse. It is, of course, true that after oil was found, access to health care and education in the Gulf did improve. And a few countries, such as Bahrain and the United Arab Emirates, have tried to diversify their economies. But booming populations in places like Saudi Arabia are eroding those gains, especially since profligate government spending, corruption, and a bloated welfare state have soaked up much of the region's historical oil windfall. The geological cost of lifting a barrel of oil out of the ground in Saudi Arabia is less than $2 per barrel, but political economists argue that all those welfare-state expenses mean that the true "social cost" may be as high as $30 or $40 a barrel. When prices fall below that level on the world market, those petro-economies falter.

What is more, a study by the IMF's Subramanian suggests that the Gulf's oil has in fact rotted democratic institutions. But surely there was never any democracy in that region to begin with, you may say. Actually, there was, of a sort. Rachel Bronson of the Council on Foreign Relations, a think tank in New York, points to life before oil as evidence. When the Saudi ruling family needed tax revenues, it consulted the merchant classes in Jeddah, so there was some mild democratic participation. The arrival of vast oil wealth, she argues, wiped out the power of the merchants and made it easier for the royal family to quash democracy.

Another recent political argument is that resources fuel civil war. An analysis by Paul Collier of Oxford University suggests that for any given five-year period, the chance of a civil war in an African country varies from less than 1 percent in countries without resource wealth to nearly 25 percent in those with such riches. That astonishing correlation goes a long way toward explaining the tragic conflicts that have torn apart Sierra Leone, Liberia, Congo, and other African countries in recent years. Take these factors together, and it becomes clear that oil fuels conflict, weakens democratic institutions, and strengthens the hand of tyrants and thugs.

Subramanian concludes that economic factors like Dutch disease and corruption alone do not explain the Oil Curse. He maintains that the problem is weak institutions. George Soros, a billionaire financier who has set up nongovernmental organizations to work on this issue, agrees heartily. Happily, international initiatives are starting to shine a cold light on the murky business of oil. Tony Blair, toward the end of his tenure as Britain's prime minister, helped start a voluntary effort involving governments and oil majors known as the Extractive Industries Transparency Initiative (EITI). George Soros is backing the Publish What You Pay campaign, which demands more aggressive disclosure of oil finances. Even some big oil companies, long accused by activists of propping up dictators with bribes, are joining the transparency bandwagon.

As with the issue of global warming, BP is the oil major making the most public noise—and ExxonMobil apparently the one most opposed to change. Ask senior executives at both firms what they think of the Oil Curse, and their answers are strikingly different. Graham Baxter at BP says, "The curse of oil is a problem that BP recognizes, and we have a part to play in helping our hosts deal with this wall of dollar-denominated cash coming into their fragile economies." But André Madec of Exxon says, "We don't like to call it the oil curse, we prefer 'governance curse.' We are private investors, and it is not our role to tell governments how to spend their money." Yet the two firms' actions are not so different. BP may be vocal on the issue, but after getting burned in Angola (it published information about its oil bid and got a bitter rebuke from government officials), it no longer strays far from the pack. And Exxon, for all its gruff talk, is at the heart of a controversial but innovative project aimed at monitoring the revenues generated by the new Chad-Cameroon oil pipeline. The money is deposited in offshore escrow accounts and scrutinized by an oversight committee representing parliamentary and civic organizations before it reaches Chad's dictator. It is also involved in half a dozen countries participating in the EITI.

How far can transparency go? Jeffrey Sachs sees no reason why government oil contracts should stay secret. Companies and governments have usually engaged in a conspiracy of silence about contractual terms, signing bonuses, and other rip-offs. But things are changing. The Western oil majors (though not state-run Goliaths in China and India) are coming to see more transparency as inevitable or even desirable. As Madec puts it, ExxonMobil wants oil revenues to "go to the people rather than accounts in Switzerland" as it helps secure his firm's "license to operate." In other words, it reduces the risk of boycotts and bad publicity.

Even the World Bank now seems keen on pushing for transparency. Despite a setback in Chad, whose thuggish government backtracked in 2006 on its promises, officials at the agency vow they will press ahead. "Countries have no justification for secrecy," insists Rashad Kaldany of the bank's International Finance Corporation. "All of these agreements will be made public in future." And the IMF is already leading the charge: it required Equatorial Guinea, Angola, and other recalcitrant countries to open up their oil accounts or risk ostracism. The good news is that some poor-country leaders are coming to the view that transparency is best. The fledgling oil states of São Tomé and Príncipe and East Timor are eager participants in the EITI. Nearly two dozen in all have joined up. Nigeria's recent participation is encouraging the rest of west Africa, Soros argues, just as Azerbaijan's involvement has shamed Kazakhstan and other neighbors into cleaning up their act. The bad news, however, is the rise in oil prices since 2000, which has led many other leaders to grab the easy money, consequences be damned. Nowhere is this more obvious, and potentially tragic, than in Venezuela—the most marvelous of countries, and yet also the most cursed by oil.

The Devil's Excrement

"I call petroleum the devil's excrement. It brings trouble. . . . Look at this *locura*—waste, corruption, consumption, our public services falling apart. And debt, debt we shall have for years." Those powerful words were uttered by Juan Pablo Pérez Alfonso, a Venezuelan founder of OPEC, during the heady oil boom of the mid-1970s. At the time, he was seen as an alarmist. In fact, he was astonishingly prescient. Oil producers vastly expanded domestic spending, mostly on lavish infrastructure projects that set inflation soaring and left mountains of debt. Worse, this did little for the poor. Venezuela has earned over $600 billion in oil revenues since the mid-1970s, but the real income per person of Pérez's compatriots fell by 15 percent in the decade after he expressed his disgust. The picture is similar in many OPEC countries. So bloated were their budgets that when oil prices fell to around $10 a barrel in 1999, a number of countries—including even Saudi Arabia, the kingpin of oil—were pushed to the brink of bankruptcy.

Oil prices rebounded, of course, and so they were saved. After several years of strong prices, OPEC is once again rolling in money. The cartel's revenues in 2006 were three to four times as high as the $120 billion it earned in 1998. However long it lasts, there remains the risk that this windfall will once again be squandered. There are signs that some countries in the Persian Gulf are trying to act more wisely than in the past. Qatar has started to diversify its gas riches into media (its emir started the Al Jazeera television network) and finance. Thanks to the foresight of the leaders of the United Arab Emirates, the sparkling city of Dubai is well on its way to becoming the next Singapore. But some of the giants of the oil world are backsliding badly. Hugo Chavez's populist regime has helped the poor a bit by redirecting some oil revenues to social spending, but studies show that poverty has been cut much less than it should have been, given the country's windfall. That is because Chavez has ignored the prescient warnings issued by his countryman, Juan Pablo Pérez Alfonso, and has been spending huge amounts of money on foreign policy misadventures and other questionable (some would say political) ventures. Vladimir Putin's clumsy and legally questionable nationalization of Russia's oil and gas sector is bad news for foreign investors, of course, but the World Bank issued a warning that his policies risked pushing the country deeper into the Oil Curse as well.

The only surefire way out of this trap is to diversify the economy. This will not be easy, but it can be done. Mexico, which embraced liberal economic reforms with gusto two decades ago, has managed to reduce oil's share of its national output dramatically, as manufacturing and other nonoil sectors have taken off. The North American Free Trade Agreement (NAFTA) helped Mexico in this regard by locking in and accelerating the reforms aimed at diversifying the economy. Could this really happen in a petro-state like Saudi Arabia? Not for ages, but there are positive signs. The Saudi rulers have opened up some sectors, such as utilities and petrochemicals, to foreign and private investors. They even encouraged diversification and liberalization of the protected domestic economy to bolster the nation's bid to join the World Trade Organization (WTO). To understand how rocky the road ahead might be, though, consider a factor that was one of the obstacles to Saudi Arabia's WTO membership: the huge subsidies paid to powerful families growing wheat, one of the most water-intensive crops in the world, in the middle of the Saudi desert. Only when such grotesque abuses end will oil wealth stop being cursed as diabolical and start being seen for what it should have been all along—a blessing.

What If?

The third great danger that petro-dependency exposes the world to is the prospect of an oil-price shock. How serious is the threat? Consider the analysis offered by someone in a good position to know, given that he and his organization are now working mightily to push the world economy into precisely such a shock.

This well-known international figure penned a heartfelt speech a few years ago he called his "Letter to the American People." In it, he said, "You steal our wealth and oil at paltry prices because of your international influence and military threats. This theft is indeed the biggest theft ever witnessed by mankind in the history of the world." The author was Osama bin Laden. The impact of those chilling words are certainly felt in today's chaotic energy markets. Oil prices shot up from $10 a barrel in 1999 to well over $40 a barrel for much of this decade. Politicians in oil-consuming economies are up in arms. Saudi Arabia, the head of OPEC, often promises relief but appears helpless. This decade's global economic boom has simply sucked global inventories dry. Nearly every OPEC producer, save Saudi Arabia, has been producing about as much oil as it can. That means that any new OPEC promise of oil will, in the future, have to come chiefly from the Saudis themselves—and that is not good news.

The main reason for worry is the so-called fear premium. Oil traders report that fears of terrorist attacks that might disrupt Middle Eastern oil exports may account for as much as $10 to $20 of the per-barrel price. That may be because what was once unthinkable now seems possible, perhaps even inevitable: a major terrorist attack, or series of attacks, on oil facilities within Saudi Arabia. For the terror premium to be justified, one needs to consider three questions: Is Saudi Arabia really so important? Would it really be possible to pull off a serious attack inside the desert kingdom? And even if such an attack were to take place, would the oil markets suffer so badly?

Until the recent rise in prices, most headlines had trumpeted the growing importance of other oil producers. The revival of Russia, overtaking even Saudi output, was supposed to undermine OPEC. Oil from Alaska would give America "energy independence." The quest for oil in the Caspian Sea was called the "Great Game." Striking oil in the waters off Brazil and west Africa was even likened to hunting elephants. Surely all this investment and discovery prove that the Saudis are ever less important to the oil market these days? Not so. Ignore the headlines and look instead at geological and market realities, and it quickly becomes clear that Saudi Arabia remains the indispensable nation of oil. The Saudis not only export more oil than anyone else but also have more reserves than anyone else—by a long shot. Nearly one-quarter of the world's proven reserves lie in Saudi Arabia. Four neighbors—Iran, Iraq, the United Arab Emirates, and Kuwait—each have about one-tenth. The reserves of Russia, Nigeria, and Alaska put together do not match Saudi reserves.

Even more important is Saudi Arabia's role as a swing producer. Unlike other countries, the Saudis have long kept several million barrels per day of idle capacity on hand for emergencies. Today, Saudi Arabia is the only country with much spare capacity available, though the precise amount is a matter of intense debate. This spare capacity allows the Saudis to moderate oil-price spikes. They have done precisely this at various times: during the Iran-Iraq War, when output from both countries was disrupted; during and after the first Gulf war, when output from Iraq and Kuwait was lost; and in 2003, when civil strife in Venezuela and Nigeria curbed output from both countries on the eve of that year's invasion of Iraq (which itself disrupted Iraqi output).

This will no doubt come as a surprise to many Americans, who are used to the demonization of the Saudis post–September 11. By acting swiftly and silently to calm the world oil markets time and time again, the Saudis have undoubtedly prevented several oil shocks and kept up their end of the Axis of Oil bargain. Whether they can or will do so in future, however, is the great question hanging over the world economy.

The Saudis say they remain keen to moderate prices by using their buffer capacity. The problem is that the more they produce in an overheated market, the less spare capacity that leaves the Saudis to prevent a further oil shock. That brings us to the second question: how vulnerable is Saudi Arabia's oil industry to terrorist attack? Not long ago, this desert kingdom was seen as a reliable and supremely safe source of oil. Indeed, some called it the central bank of oil. After the September 11 attacks, such assessments seem too rosy. Terrorism is now clearly a serious problem. Not only were most of the September 11 suicide attackers Saudi (fifteen out of nineteen), but reports have suggested that some Saudis even fought against American-led coalition forces in southern Iraq. And increasingly, Saudi terrorists are striking targets at home—including oil infrastructure. In 2004, one group killed several foreigners working at a petrochemical complex in Yanbu', the biggest oil-export terminal on the country's Red Sea coast.

Not everyone is worried. Nawaf Obaid, an adviser to the Saudi royal family, argued in *Jane's Intelligence Review* that the risk of a successful attack on oil facilities remains "very low." He explains, "At any one time, there are up to 30,000 guards protecting the Kingdom's oil infrastructure, while high-technology surveillance and aircraft patrols are common at the most important facilities and anti-aircraft installations defend key locations." Obaid claims that the Saudi government has added $750 million over the past few years to its security budget (which totaled $5.5 billion in 2004, according to him) specifically to fortify the oil sector.

It is true that Saudi oil infrastructure would be pretty hard for terrorists to take down. There is plenty of redundancy built into the Saudi network—through multiple ports, pipelines, and excess capacity—that should ease the blow from any attack. Besides, to do any real damage, terrorists would have to hit bottlenecks, not just blow up random bits of pipeline. Kevin Rosser of Control Risks Group, a security consultancy, quips that "the golden goose is not a sitting duck." But other security experts think that goose may yet be cooked. James Woolsey, a former head of America's Central Intelligence Agency, is unimpressed by talk of improved security: "Guards and fences are easy to put up, but they don't defend against the real threats." Trucks have to come in and out of facilities, he observes, and Aramco employees and security guards have to move about. He thinks that several attacks, if coordinated by terrorists who have infiltrated Aramco, could cripple the Saudi system.

Robert Baer, the intelligence expert who was the inspiration for George Clooney's character in *Syriana*, offers some suggestions in his disturbing recent book *Sleeping with the Devil*. He points out that Ras Tanura, a port on the Gulf, is a vulnerable terrorist target. With an output of perhaps 4.5 million barrels per day, this is the biggest oil-exporting port in the world. This single spot, the size of a few football fields, exports more oil every day than the entire OPEC countries of Indonesia, Nigeria, and Venezuela combined! Baer thinks a small submarine or a boat laden with explosives (as happened in October 2000 with the attack on the USS *Cole* off the coast of Yemen) could knock out much of Ras Tanura's output for weeks or even longer.

An even scarier possibility raised by Baer is the crashing of a hijacked airplane into Abqaiq, the world's largest oil-processing complex. If done with the help of insiders, he speculates that the facility's throughput (nearly seven million barrels per day, by his estimate) would be choked off to as little as one million barrels per day for two months—and might remain as low as three million for seven months. Woolsey adds that an attack using weapons of mass destruction (especially "dirty bombs") would be even more devastating than one that used mere airplanes. That's because the special equipment and safety precautions required for such a cleanup would greatly delay the restart of production. All told, the pessimists reckon that well-coordinated attacks could take as much as six million to seven million barrels per day of Saudi output off the market for weeks—and possibly much longer, if dirty bombs were used.

The world is clearly better equipped to handle a supply shock, even one caused by terrorists, than it was during the turbulent 1970s. For a start, the rich world is much less energy intensive. Unlike three decades ago, America, Europe, and Japan now maintain large "strategic reserves" of petroleum and coordinate the release of these during emergencies through the International Energy Agency (IEA). The rise of energy futures markets over the past two decades also offers some scope for the world to deal better with short-term price shocks, since risk-averse parties can lock in prices through long-term hedging contracts.

Yet there is cause for concern. The unprecedented prospect of Saudi Arabia being under attack from within exposes the vulnerabilities of the world's two chief forms of insurance against oil shocks: IEA stocks and Saudi swing capacity. One problem is that the world's strategic petroleum stocks (which are stored in such places as salt domes in Louisiana) simply cannot be drawn down all at once. If a loss of Saudi output is anything like as long-lived as Baer fears, or if other Middle Eastern output is also lost at the same time, then strategic stocks may prove inadequate. Prices will soar, and the market will return to equilibrium only through painful cuts in consumption and accompanying losses in economic output and welfare.

The more troubling revelation surrounds Saudi swing capacity. There is the obvious point, of course, that this particular insurance policy will not be worth very much if there is a serious supply crisis inside the kingdom. However, even if the horror scenarios never happen, the global spare-capacity crunch is still alarming. Amy Jaffe of the Baker Institute, at Rice University in Texas, observes that in 1985, OPEC maintained about fifteen million barrels per day of spare capacity—about one-quarter of world demand at that time. In 1990, when Iraq invaded Kuwait, OPEC still had about 5.5 million barrels per day of spare capacity (about 8 percent of world demand). That, she argues, meant that the cartel could easily and quickly expand output to absorb several disruptions at once.

That is no longer true, and the world's safety net now looks threadbare. Today's spare capacity of around two million barrels per day is less than 3 percent of global oil demand—and it is almost entirely in Saudi hands. And yet, "normal" threats to supply that fall far short of doomsday terrorist scenarios remain. Venezuela, Indonesia, Nigeria, and other oil-rich states face frequent political tests that could cripple oil exports. As the *New York Times* showed in 2006, the terrorist threat to Iraq's oil infrastructure morphed from ad hoc attacks to systematic disruption caused by a powerful oil-smuggling mafia. Whether the Saudis can handle the consequences of all this is entirely unclear.

And yet somehow they must. There is little chance that the Saudis will be dislodged as the swing producer anytime soon, even by a resurgent Iraq. One reason is cost: the Saudis sit atop the cheapest reserves in the world. Another is the fact that Saudi oil remains in state hands. Aramco does not have to justify to shareholders the billions it wastes on idle fields—a luxury that Western oil majors do not enjoy. Even if Iraqi oil remains in state hands, as appears likely, a cash-starved, independent Iraqi government could not justify developing fields that will remain idle. So long as the world remains addicted to oil, it is hopelessly dependent on Aramco, the Goliath of the oil world, defending the entire global petro-economy.

A Witch's Brew

Of course, disaster might not happen. The Middle East may cool down, Osama bin Laden may be caught and al-Qaeda disrupted; oil-guzzling in America and China may yet slow down. Oil prices have already shown signs of moderating; they might even collapse as they did in the mid-1980s and again in the late 1990s. If they do, however, it would be wise to remember just how precariously the world's oil markets have been balanced of late—and may be again in the future, if things take a wrong turn in Saudi Arabia.

A witch's brew of soaring oil demand, private-sector de-stocking, and lack of investment in new production capacity by OPEC has left the world with an extraordinarily tight oil market. There is less spare capacity than at almost any point in the past thirty years. As Edward Morse of Lehman Brothers, an investment bank, puts it, "The world has been living off surplus capacity built a generation ago, and thought it could get by. It turns out not to be the case." Building a new surplus will inevitably take a long time. Until then, the potential instability of Saudi Arabia's oil supply will remain a strategic weakness for the world economy.

All the hardware and military might in the world cannot provide 100 percent certainty that the nightmare will not come to pass: a disgruntled employee sneaking a dirty bomb into Ras Tanura, the world's largest oil-export terminal, or Abqaiq, the vital oil-processing center. Ras Tanura has already been attacked figuratively by radical Islamists disguised as local fishermen in *Syriana*'s spectacular closing sequence, but could that happen in real life too? That was the question that struck one of the authors while gazing out from the port captain's panoramic watchtower overlooking Ras Tanura. The waters around the port complex were teeming with tiny fishing vessels from nearby villages, vessels that looked an awful lot like the little fishing boat in *Syriana* used by terrorists in the grand finale to blow up a vital port. Officials on hand pointed to all of the high-tech security they use to keep the terminal safe, from aerial and marine patrols to redundant computer systems to multiple rings of fencing and multiple layers of security forces.

And what about all of those tiny fishing boats darting in and out of the oil-loading area—presumably they are tracked very carefully by radar and GPS? Well, not really, was the embarrassed response from officials in the watchtower. The boats, they said, were too small to show up on the watchtower's monitoring equipment. "Don't worry, we know who each and every one of those fishermen is," was the less-than-reassuring reply from the oil minister, Ali Naimi, when confronted with this gaping hole in security. Even less reassuring was news of an attack by heavily armed fighters on Abqaiq in early 2006, foiled in the nick of time by Saudi authorities.

Given all these vulnerabilities, it must be reassuring for the Saudi regime to know that the world's greatest superpower remains committed to its defense, thanks to the oil-for-security pact forged by FDR and King Ibn Sa'ūd. When that notion was put to Ali Naimi at his ministerial headquarters in Riyadh, he gave a stunning response. Shaking his head, he started by saying, "That is really a topic for the foreign minister." As a wry smile crossed his face, he continued, "But given what has happened after September 11, I am not so sure that we can count on the Americans to hold up their end of the bargain."

That is somewhat worrying: the Saudi oil minister, the guardian of the oily end of the Axis of Oil, admitting his personal doubts about the reliability of his country's American allies. If you thought the world's addiction to oil was dangerous when the Saudi-American alliance was strong, just imagine how precarious things really are if senior Saudi figures feel they cannot fully trust the Americans to come to their defense if they are under attack. Thanks to oil's unavoidable geopolitical tangles, the world looks to become an ever more dangerous place.

Ali Naimi's haunting words were still fresh in one of the author's minds as he went to dinner in Riyadh that same evening with two leading security experts. The venue was the panoramic restaurant high up in the Al Faisaliah skyscraper, whose strikingly original design makes it look like that eerie evil-eye erection in the movie version of *Lord of the Rings*. Even as Nawaf Obaid was arguing that the Saudis had broken the back of the al-Qaeda cell in his country, a massive explosion went off nearby. He continued apace, oblivious to the fact that others at the table were staring out the skyscraper's panoramic windows at helicopter gunships flying at eye level behind his back. Then another explosion went off. Two car bombs went off that evening within line of sight of the supper table.

Oil prices jumped on cue, then fell again once markets realized the attackers had failed to blow up their intended targets. But if those suicide bombers had successfully struck where al-Qaeda now says it wants them to, oil prices would have surged to $150 a barrel or more. That, after all, is exactly what Osama bin Laden has indicated very clearly he wants. He has even made clear what is the "right" price for a barrel of oil: $144. Several years ago, the leader of the al-Qaeda terrorists issued a little-noticed proclamation on energy economics. In it, he accused the United States of "the biggest theft in history" for using its military presence in Saudi Arabia to keep oil prices down. In his view, that larceny added up to $36 trillion. America, he insisted, now owes each Muslim in the world around $30,000, and counting. Six decades after it was first forged, the Axis of Oil is in trouble.

Given that even the Saudis are questioning its relevance today, surely it is time for the U.S. to do the same. After all, how can America's leaders possibly ignore the threat to the world economy arising from oil when Osama bin Laden's warnings have been so clear? Unfortunately, as the next chapter on global warming shows, Washington, D.C., has been ignoring equally clear warnings about the dangers arising from climate change—until now. Welcome to the Great Awakening, which could grow to be the most important political force of this new century.

CHAPTER SIX

The Slumbering Giant Awakes

The Great Awakening of America to the dangers of oil addiction and global warming is pushing corporations to act—but can big business really solve the problem?

One spring evening in 2006, the ancient but brilliant Snooks Eaglin played blues guitar at the Rock 'n' Bowl. A joyous crowd of New Orleans locals bowled and flirted and drank and danced, and the smell of gumbo and crawfish filled the air. In that place, with that crowd, on that night, it was easy to believe that the Big Easy was back. The legendary nightspot, famous for offering bowling and zydeco bands under the same roof, was packed to the rafters. "Rock 'n' Bowl will never die!" was the club's long-standing motto—and, just like the hubris that comes so naturally to the region's oil industry, that motto seemed to capture the city's indomitable sense of self.

But New Orleans is not indestructible, as everyone knows today. The short drive from the French Quarter out to the Rock 'n' Bowl revealed block after block of devastated homes and blacked-out neighborhoods. An analysis done by the Brookings

Institution revealed that nearly a year after Katrina hit, fewer than two-thirds of the city's homes had electricity, only half of the city's hospitals, and fewer than one in five of its public schools were reopened. Even the strip mall that Rock 'n' Bowl calls home sat empty. In short, conditions in much of the city were still appalling.

That is bad enough, but the really disgraceful part of the Katrina story is that this tragedy could and should have been averted. New Orleans has been living on borrowed time since 1718. That was the year Jean-Baptiste Le Moyne, sieur de Bienville, decided to build a French settlement on a hurricane-infested, low-lying swath of swampland surrounded by three bodies of water. Since then, experts have repeatedly issued warnings about the city's vulnerability to intense storms and flooding, only to be ignored time and again—just as those who warn of the risk of a politically induced oil shock have also been ignored.

Environmentalists had cautioned that the federal government's tampering with the Mississippi River to help the shipping and oil industries had destroyed so much coastal marshland that the city's natural flood defenses were dangerously eroded. And climate scientists, who had observed the waters of the Gulf of Mexico getting hotter, had argued that climate change could well lead to much more intense and destructive hurricanes. Even an expert from the Army Corps of Engineers, which is partly responsible for the city's levee failures, urged officials before the storm hit to shore up their defenses: "You're living on borrowed time today. You have until the next big storm zeroes in on coastal Louisiana directly." And yet leaders took no action.

When the double whammy of hurricanes (Katrina and soon thereafter Hurricane Rita) hit in 2005, it battered not just the communities of the Gulf Coast but the region's oil infrastructure too. No hurricane season had ever destroyed as much of the entire value chain of energy infrastructure, from offshore rigs to underwater pipelines, to refineries and power lines onshore. Worse yet, this coincided with a national shortage of gasoline. Perhaps half of the output of the Gulf of Mexico, itself accounting for about a quarter of America's oil and gas output, was shut down in the wake of the two storms. So serious was the crunch in refining and refined products such as gasoline that the world's developed countries agreed to something they had never before done in peacetime: an emergency release of government strategic stocks of crude oil and gasoline to reduce the possibility of an oil-price shock.

They were worried because the world's refining sector was already stretched to the limit before Katrina hit. The problem was most acute in America, where demand for gasoline has been far outpacing refining capacity in recent years. As a result, America imports about a tenth of its gasoline, up from just 4 percent a decade or so ago. Domestic refining volumes have gone up, but not enough; and Katrina knocked out eight refineries, representing perhaps a tenth of the country's capacity. Rita came along in its wake and actually did far more damage than Katrina to offshore infrastructure: dozens of rigs and platforms were damaged, and miles of underwater pipelines were ripped out and knotted up like pretzels.

Refinery workers were unable to return to their jobs for weeks, as most lost their homes, and some lost family members. In the wake of the storms, gasoline prices in some areas jumped a dollar, touching $4 a gallon or higher. In other places, shortages and panic buying were reported. With most commodities, higher prices dampen demand. The problem is that demand for gasoline is inelastic in the short term. Even if prices rise overnight, frazzled parents still have to go to work and drop off their children at soccer practice.

For the first time since the 1970s oil shocks, there was semiserious talk in official Washington of conservation measures. The full Senate Committee on Energy and Natural Resources sent a letter to George Bush asking him, among other suggestions, to encourage federal government employees to share cars. Senator Pete Domenici, the Republican head of the committee, even floated an idea that Bush never liked: tightening regulations that would force an improvement in vehicle fuel-economy levels, which were close to a twenty-year low.

The great unknown about oil was whether the world would see a second big blow to the energy system before the damage done by the twin terrors of Katrina and Rita was undone. This question was all the more urgent, since governments had committed themselves to releasing a chunk of their official stocks of petroleum onto the open market. Once those were depleted, the world would obviously be much less able to cope with all the other common surprises (coups, civil strife, or terrorist attacks) that lead to supply disruptions in oil-producing countries.

What would have happened to the world economy if there had been another big blow to the oil industry posthurricanes—or if, in the future, two moderate blows buffet the world oil economy at once? America's National Commission on Energy Policy, a group of leading energy gurus and politicians from both parties, asked that question before the storms. In an elaborate exercise, the group "gamed" out the likely impacts of a natural calamity and a supply disruption amounting to just 4 percent of the world's oil supply. Their conclusion: oil prices would probably jump from $60 per barrel (their assumed baseline) to over $160 per barrel, even without any massive al-Qaeda attack on vital Saudi oil infrastructure. With that kind of price shock, the world economy would surely be pushed into the worst recession since the oil shocks of the 1970s.

At the time of the first Gulf war, such a prognosis could never have been made. Back then, global demand was much lower, and the world had plenty of spare production capacity. But today, as the run-up in prices since 2000 has already indicated, there is simply no longer any safety net in the global petroleum system. And as Katrina made painfully obvious, adding the new dangers posed by global warming to the mix leaves the world even more vulnerable to an oil shock.

Indeed, that is just about the only positive thing to emerge from this tragedy: Katrina served as a wake-up call for the United States on global warming. Before the storm, opinion polls showed Americans were oddly out of step with the rest of the world on the issue of climate change. Many were willing to believe self-serving claims dished out by the oil industry and the Bush administration that climate science is too "uncertain" to merit serious action. That was before Katrina.

After the Storm

"The greatest threat of climate change for human beings, I believe, lies in the potential destabilization of the massive ice sheets in Greenland and Antarctica. As with the extinction of species, the disintegration of ice sheets is irreversible for practical purposes. Our children, grandchildren, and many more generations will bear the consequences of choices that we make in the next few years." With those powerful words in the *New York Review of Books*, Jim Hansen, the head of the Goddard Institute for Space Studies and NASA's chief climate scientist, made clear to the world that he would not be silenced by his political masters.

Before Katrina struck, the Bush administration had repeatedly tried to muzzle or discredit climate experts like Hansen, who spoke out about the dangers posed by climate change. When the UN's Intergovernmental Panel on Climate Change, a collection of thousands of leading global experts on climate, issued an authoritative report on global warming at the turn of the century, officials in the Bush administration scoffed that they were foreigners. President Bush set up his own handpicked group of American climate experts at the National Academy of Sciences in 2001 to look again at the question. To his dismay, this team of experts reaffirmed what the UN panel had said:

> *Greenhouse gases are accumulating in Earth's atmosphere as a result of human activities, causing surface air temperatures and subsurface ocean temperatures to rise. Temperatures are, in fact, rising. . . . Human-induced warming and associated sea level rises are expected to continue through the 21st century.*

Humiliatingly for Bush, his handpicked, all-American experts confirmed that climate change is real, potentially devastating, and partly man-made. And still, Bush chose to brush aside their recommendations. It even surfaced that administration officials friendly with the oil industry censored scientific reports to tone down the calls for action. When one such case became public, the official in question was forced to leave the White House in disgrace—only to wind up immediately on the Exxon payroll. But the Bush team finally met its match in Jim Hansen. NASA's leading climate guru had never been one to conform or be swayed by political pressure.

Though he was one of the first scientists to sound the alarm bells on man-made global warming, he never became an extremist. When Bill Clinton asked him back in the early 1990s to write a rebuttal to a *New York Times* Op-Ed article that played down global warming and criticized Vice President Al Gore, he refused. To the annoyance of environmentalists keen to promote renewable energy as the solution to global warming, Hansen wrote a paper making clear that cutting emissions of methane (which is a potent greenhouse gas) worldwide would be an even better way to fight climate change right now, since it would provide more bang for the buck than putting up expensive windmills.

And when the Bush administration's press handlers tried to censor his speeches and suppress his views, Hansen refused to bow. He went public with his grievances around the time Katrina hit. He became something of a media sensation for having the courage to do what dozens of other federal bureaucrats were too afraid to do: speak out against the Bush administration's suppression of "inconvenient" climate science. In doing so, Jim Hansen played a vital role in America's Great Awakening to climate change.

In the wake of Katrina, America at large finally started paying attention to global warming. In a way, this is unscientific: no single storm, even one as devastating as Katrina, can conclusively be linked to global warming. Rather, all careful scientists will say is that warmer, more energetic oceans and more intense, perhaps more frequent hurricanes are consistent with what they expect with global warming.

Such nuances did nothing to deter the veritable flood of media coverage of the topic, however. Many television shows, ranging from CNN shock-and-awe programs to thoughtful Tom Brokaw investigations to alarmist specials even on the Bush-friendly Fox News Network, talked up the dangers. *Vanity Fair* ran a special green issue in early 2006, calling global warming "a threat even graver than terrorism." Picking up on Hansen's fears of a polar meltdown (a remote but nevertheless frightening possibility), the magazine's photo editors ran pictures of what Washington, D.C., San Francisco, Manhattan, and other places would look like in a global-warming nightmare. Seeing pictures of their overpriced beach houses in the Hamptons and Martha's Vineyard underwater finally got even the most self-absorbed and ungreen to pay attention to this issue.

And it was not just the elites who were jolted awake by Katrina. *Newsweek* ran its own cover story on the matter in July 2006, "The New Greening of America," which argued that the country as a whole has regained an environmental consciousness not seen since the early days of the modern environmental movement three decades ago. Membership in the Sierra Club had shot up by a third since 2002. A backlash against George Bush's poor environmental record was undoubtedly one factor, but as the club's executive director, Carl Pope, told *Newsweek*, so too was Katrina, because it "changed people's perceptions of what was at stake." GlobeScan, a leading international environmental polling outfit, confirmed the hurricane's impact: Americans had consistently ranked global warming lower as a concern than did respondents from other parts of the world—until 2006. But in an important poll taken before the media hoopla surrounding the launch of *An Inconvenient Truth*, Al Gore's movie and book on climate, Americans told GlobeScan they are now just as concerned about global warming as are Europeans or Asians.

So what should America do now that it has woken up to the perils of global warming? Of course coal is a big part of the problem, but it is revealing that Hansen pointed to the even bigger but little-publicized carbon nightmare of "manufacturing" gasoline from unconventional petroleum. He put it this way in his call to arms in the *New York Review*: "We have at most ten years—not ten years to decide upon action, but ten years to alter fundamentally the trajectory of global greenhouse emissions. . . . If instead we follow an energy-intensive path of squeezing liquid fuels from tar sands, shale oil, and heavy oil, and do so without capturing and sequestering carbon-dioxide emissions, climate disasters will become unavoidable."

Oil and cars are really the thorniest part of the climate puzzle. While burning coal is more carbon-intensive than burning conventional oil, electricity emissions are far easier to tackle than those from transport. One reason for that is the fact that coal-fired electricity has many proven and powerful substitutes, in the shape of windmills and gas turbines and nuclear plants; in contrast, oil still retains a near monopoly grip on the world's car and bus fleet. What is more, if push comes to shove, politicians can round up the several dozen bosses of coal-fired utilities in America, put a figurative gun to their heads, and force them to use clean coal technologies or switch to low-carbon alternatives. There is no plausible way of convincing 300 million Americans to part with their beloved but carbon-spewing automobiles. Another factor is the lag time for new automotive-emission technologies to make a difference, given that the average car lasts fifteen years on the road. The only way the world can hope to tackle the carbon challenge posed by cars and oil is if policy makers and industry leaders show genuine leadership. This is particularly true in the United States, by far the biggest market for cars and oil and therefore the one that sets the global trends. So are America's political and industrial leaders up to the task?

While Rome Burns . . .

"Conservation may be a sign of personal virtue, but it is not a sufficient basis, all by itself, for a sound, comprehensive energy policy." So declared Dick Cheney, America's vice president, in

2001, as he defended his administration's then-new energy policy. The administration's aim is to bolster energy independence from OPEC by boosting domestic supplies of energy, including oil found in protected parts of Alaska.

Alas, America will never achieve energy "independence," given that it consumes a quarter of the world's daily oil production but sits atop less than 3 percent of its proven reserves. All the oil in all the protected lands in the country would not make a darned bit of difference in the global equation. And yet that did not stop Congress from invoking "energy independence" to justify passing Cheney's energy proposals in the form of an energy bill in 2005. That law will hand out, on one independent estimate, over $80 billion in pork to the energy industry—with the lion's share going to profitable, well-capitalized incumbents in the oil, gas, and nuclear businesses that should not receive even one penny of subsidy.

The obvious culprits are oil and car lobbyists and venal politicians, of course, but the surprise is that environmental lobbyists are also part of the problem. Rather than demand an end to all subsidies and a level playing field for clean energy, most of them just begged for a few crumbs for their pet wind, solar, or efficiency projects. That sort of special-interest lobbying in effect provided political cover for the much bigger subsidies provided for hydrocarbons.

It would be easy to blame this fiasco on Bush and Cheney, two cronies of the oil and car industries—but that would let too many other culprits off the hook. As with the militarization of America's energy policy abroad, the pork-barrel approach to energy policy at home is very much a bipartisan affair. Republicans controlled both chambers, it is true, but the 2005 energy bill that passed was strikingly similar to one crafted earlier by the Democrats when they controlled the Senate. In the end, the horrid bill became law with overwhelming support from both parties.

Even more striking has been the bipartisan opposition to meaningful action on climate change in Washington, D.C. (a trend curiously contradicted by the bipartisan push for climate action from the governors of both "red" and "blue" states). One of the most useful measures the government could take to tackle both climate change and oil addiction, most economists agree, is to impose a tax on the carbon content of fuels. Bill Clinton courageously proposed an energy tax along those lines in 1993, but he could not get it through a Congress packed at the time with friendly Democrats. Al Gore negotiated the Kyoto Protocol on America's behalf, but his administration did not dare present the pact to the Senate for ratification, as virtually every Democrat would have voted against it.

More distressingly, the Clinton administration did not even try to push any tough carbon legislation through Congress during the 1990s. Yes, they would likely have lost the votes, but they could have shamed the obstructionists and industry cronies into facing media scrutiny and the wrath of voters. That decade of policy inaction on global warming, combined with the roaring miracle economy of the 1990s, led to America's greenhouse-gas emissions going up by around a quarter during that decade. This gassy expansion made it prohibitively costly for any American president in 2000 to try to meet Kyoto targets. It also made it needlessly easy for Bush to yank America out of the UN treaty, which he did in early 2001.

This potted history of climate politics in Washington explains why Europeans are wrong to blame the "Toxic Texan," as London's *Independent* newspaper dubbed Bush, for America's inaction on climate. American politicians of both parties, responding in part to apathy among ordinary voters on the issue and not just to special-interest lobbying, simply did not see the issue as a clear and present danger—that is, until Katrina hit. And yet, across the Atlantic, one government after another, regardless of political persuasion, has rushed to act against global warming. In 2006, Britain even saw the bizarre spectacle of the leader of the Conservative Party trying to outdo Prime Minister Tony Blair by demanding even more aggressive action on climate change. Just try to imagine Trent Lott or Newt Gingrich straining to outgreen Al Gore, and you will get an idea of how seriously Europeans take climate change.

Washington's Oil Curse

Since George Bush is not a lone gunman, what explains the American exceptionalism on climate change? There are some legitimate factors at work that explain why America was so slow to wake up to the dangers of global warming. For a start, boosted by parliamentary voting systems that grant minority parties a lot of power, green parties are influential in many parts of Europe. In some countries, such as Germany, they have even held the environment portfolio in the cabinet and so have been able to push the European Union to act much more aggressively than any consensus that could emerge from America's two-party system.

Another factor at work is the widespread fear in Europe that it simply has more to lose from dangerous levels of climate change than America does. The United States is a huge, continental nation, spreading from the tropics to the tundra, from marshlands to deserts. Western Europe, on the other hand, is a small and vulnerable parcel of land situated quite far north, into the upper reaches of the Northern Hemisphere. If one went strictly by latitude, France and Britain should have winters as frigid and joyless as those endured in, say, Alberta. However, because of the Gulf Stream ocean current, which circulates warm water from the Caribbean up to the western shores of Europe, the Old Continent has rather mild winters and pleasant summers. There was widespread fear that the influx of freshwater from melting polar ice could shut down the Gulf Stream.

Carl Wunsch of MIT has shown that this so-called conveyor belt is actually driven mostly by the prevailing winds caused by the earth's rotation, not the salinity differences between warm and cool waters. So the shutting down of this current because of melting ice and its lower salinity is one climate nightmare that may well be overplayed. Nevertheless, a very loud chorus of doom-saying environmentalists has convinced the European media and body politic that global warming could lead to this circulation pattern collapsing—and so plunging their continent into a frigid and bleak future. Self-interest certainly sharpens one's focus on a problem.

That points to another, more shameful part of the explanation for why Washington has not acted on climate. Historically, America has been home to the most important companies in the car and oil industries. The big bosses of Detroit and especially of Texas have been paranoid that their short-term business interests will be hurt if Washington moves to tackle global warming since the use of their products is at the heart of the problem. While there are oil and car companies in other parts of the world, nowhere is their lobbying power and cultural resonance as great as in America, though Mercedes and BMW in Germany have come close. But, if that vested interest could be overcome, what would it make sense for America to do about global warming?

CAFE Culture

A good first step for American climate policy would be to tackle efficiency. Alas, in keeping with Cheney's dismissive views on conservation, there is little in the 2005 energy plan (which Congress passed just before Katrina hit) to encourage greater fuel economy in cars or gas-guzzling SUVs. That is a pity, for history shows that conservation can be a powerful check on the OPEC cartel. After the oil shocks of the 1970s, the developed world embraced aggressive policies to encourage energy efficiency. In Europe and Japan, these took the form of energy taxes; America chose instead to regulate the auto industry through the Corporate Average Fuel Economy (CAFE) law.

At the time, the conventional wisdom held that energy use and economic output would always grow in lockstep. Amory Lovins, head of the Rocky Mountain Institute, a natural resources consultancy, argued that there was an alternative "soft path." He was widely ridiculed, but the 1980s proved him right. Thanks to government policies, the rich world's energy use and GDP decoupled, and the world's developed countries grew much more energy efficient.

The biggest success was the CAFE law. With the help of high energy prices, that law led to an astonishing increase in the average fuel efficiency of new American-made cars of over 40 percent from 1978 to 1987. From 1977 to 1985, the volume of America's net oil imports fell by nearly half, even as its economy grew by a quarter. That is proof positive that higher fuel efficiency can go hand in hand with economic prosperity—and clearly refutes the Kyoto-bashing argument that tackling global warming will inevitably lead to economic ruin. Lovins believes this "broke OPEC's pricing power for a decade," as the world enjoyed low and stable oil prices from the late 1980s through much of the 1990s. More striking, he now argues that the world can repeat that trick once again, so great are the remaining inefficiencies in how the world uses energy.

It's Not Easy Being Green

It is fashionable in some quarters these days to argue that big business can be part of the solution to climate change, not just the source of the problem. "Corporate social responsibility" has been all the rage in boardrooms on both sides of the Atlantic, and most major companies now issue mind-numbing annual "sustainability" reports. Some even profess to believe that companies are not in business merely to earn returns for shareholders, but rather to pursue the triple bottom line, or people-profits-planet, or win-win-win, or some other faddish slogan.

Ask NASA's climate guru Jim Hansen whether big companies are today part of the climate solution, and he will deliver a dark and unequivocal answer: no. Thanks in part to his personal experiences dealing with undue corporate influence leading to the suppression of scientific findings and outright censorship, he takes a very skeptical view of big business. Companies often have two personalities, he cautions, a seemingly progressive public façade and a much more sinister and self-interested private personality. His views reflect those of environmentalists who claim that oil and car bosses who talk endlessly about going green but actually do little of substance are merely guilty of "greenwash." And right at the top of the list of candidates for that undesirable label is the great-grandson of Henry Ford.

Bill Ford, chairman of Ford, is a great Jim Henson fan. No, not the climate guru who thinks big corporations can be two-faced liars—he likes the fellow with a similar-sounding name who invented the Muppets. So much so, in fact, that Ford ran an environmental marketing campaign using Kermit the Frog as a mascot. That's not really such an odd choice when you consider that Kermit's favorite saying could easily be Bill Ford's personal mantra: "It's not easy being green."

Few doubt that Bill Ford is, at a personal level, a committed environmentalist. Indeed, there is plenty of evidence of his crunchy-granola ways going back to his college days. So when he came to the top job at his family's company promising to take sustainability seriously, he was welcomed with open arms by green groups. Chief among them was the Sierra Club, which went to unusual lengths to praise him for his vow to introduce hybrid cars rapidly and to put global warming at the top of the agenda. He was once even invited to speak at a conference run by Greenpeace, which has historically been anticorporate. At that conference in London, he uttered these fateful words: "I believe that fuel cells will finally end the hundred-year reign of the internal-combustion engine."

It all sounded too good to be true, and sadly it was. While Toyota brought hundreds of thousands of hybrid cars onto the market, Ford's offerings trickled out at a snail's pace. His vow to increase fuel economy dramatically also fell by the wayside as the company found SUVs too profitable to give up. That strategy proved yet another short-termist mistake in Detroit, for the market for gas-guzzlers collapsed after Katrina hit. Worse yet, the company's green promises suddenly looked mighty hypocritical. When Ford introduced the Excursion, a gargantuan SUV, the Sierra Club viciously attacked the model as the "Ford Valdez" (a reference to the notorious Exxon oil spill in Alaska). Bill Ford had gone from green hope to green hoax, one environmentalist complained to the *New York Times* in 2006.

In fact, Bill Ford always had the deck stacked against him. For one thing, he was seen as a lightweight within the company, holding the reins of power only by virtue of his family connection.

As a result, he never had a good grip on his firm's formidable bureaucracy. Second, as the firm's inability to steer away from making ever more SUVs showed, it is still driven by a focus on short-term profits that characterizes the American car firms. In contrast, Toyota's great green successes are the result of long-term vision and patient investment in new and therefore financially risky technologies like hybrids today (and perhaps "plug-in" hybrids and fuel cells tomorrow).

The biggest obstacle to greening Detroit, however, is surely corporate culture. After decades of perfecting internal-combustion technology, and with many tens of billions of dollars of investments sunk into that technology to defend, the big car companies are not very receptive to environmental changes that threaten existing assets—even when the orders come from the very top. GM's former chairman Roger Smith (who achieved notoriety as the subject of Michael Moore's breakthrough film *Roger and Me*) actually championed the first modern-day electric car, the superquick and supersleek EV1—only to find his efforts thwarted by his own bureaucracy. To add insult to injury, even as GM killed the EV1, it was busy putting the company's marketing might behind the most environmentally unfriendly car ever made: the Hummer. Bill Ford has run into much the same cultural problem at his company. Hans-Olov Olsson, then the head of Ford's Volvo division, explained, "When Bill Ford said he wanted to end the internal-combustion engine, thousands of engineers at the company who had spent their lives perfecting that technology simply didn't believe him."

Turning an American car company in a greener direction is clearly tricky, but that is nothing compared to the obstructionism seen at Big Oil. Throughout the 1990s, when the UN's Kyoto Protocol was being worked out, the American oil industry led efforts to kill that pact, eliminate any alternatives, and generally trash climate science altogether. Their campaign, organized under the umbrella of the Global Climate Coalition (GCC), was a formidable, well-financed, and highly successful one. Exxon-Mobil has been the principal force lobbying against action on climate—perhaps unsurprisingly, given Lee Raymond's view that climate change is a giant hoax perpetrated by government scientists. Other American oil companies do not dare speak so boldly in public, but they have quietly stood shoulder to shoulder with Exxon in the GCC and other lobbying efforts on climate.

What is striking here is that the European oil giants, led by BP, have taken a much more forward-looking line on global warming—much as the Japanese carmakers have taken a much more innovative tack than their stodgy Detroit counterparts. The leader of this pack, and a pioneer of corporate social responsibility, is BP's boss, Lord Browne.

His company's splashy advertisements proclaiming itself "Beyond Petroleum" are hard to miss. But is this a genuinely greener strategy than Exxon's—or, like Bill Ford's failed efforts at his car company, is it merely greenwash?

How Green Is Browne?

Back in 1999, Lord Browne concluded a deal that could yet revolutionize the energy business. No, not the $27 billion take-over of Atlantic Richfield (ARCO) that transformed BP into the world's second-largest oil company and hogged the headlines. The really profound decision, made around the same time, was his decision to spend $45 million to win control of Solarex. That made BP the world's biggest solar-energy company of the day.

Despite its trifling size, the deal raised interesting questions about the thinking of BP's leader. Browne's financial savvy helped turn BP into one of the world's highest-rated, least sentimental majors. He shocked the industry with his early acquisition of Amoco, setting off a round of mergers that included the marriage of Exxon and Mobil, Total and Petrofina, Chevron and Texaco, and various others. By gobbling up ARCO, he once again caught his rivals off guard. "He clearly has the first-mover advantage," said Robin West, an American industry expert, at the time, "and the industry is dancing to his tune."

Yet the man who made a name for himself as a ruthless cost-cutter and daring deal maker suddenly seemed to be behaving like a quixotic dreamer. Solarex was not the only example, for he also pushed ahead with other environmentally friendly efforts. BP now powers many of its gasoline stations worldwide using solar panels. The company has divisions devoted to wind, solar, and hydrogen energy. And in 2006, BP announced a big joint venture with DuPont, the world's second-biggest chemical company, to produce commercial quantities of biobutanol—a renewable biofuel that can be blended into existing gasoline infrastructure easier than ethanol can.

All that is fine, say eco-skeptics, but surely all oilmen are tainted by original sin and cannot truly go green. The green-wash argument about BP (and Shell, which is also investing in renewables and making green claims in its advertising) goes as follows: Yes, it's true that BP is investing on the order of $1 billion to $2 billion a year in clean energy technologies—but so what? The company, like its American rivals, is still spending a whopping $10 billion to $15 billion a year on the exploration and production of dirty oil and gas. The BP ads may say "Beyond Petroleum," but in fact its profits still come from *burning* petroleum; the firm is likely to remain a peddler of hydrocarbons for many years hence. The green backlash challenged Lord Browne's hard-won reputation as a new kind of energy boss.

As if the tarnishing of his green credentials were not enough, oil's most celebrated chief executive fell even further from grace in 2006. An explosion at one of its refineries in Texas and evidence of corrosion in its Alaskan oil pipeline network propelled the firm into the headlines. The boss who seemed to have the golden touch suddenly seemed vulnerable. An investigation conducted by James Baker, the former secretary of state, pinned the blame for the Texas accident on various BP executives for ignoring safety. That was a fair assessment, and BP responded by admitting responsibility and firing senior managers in charge. But the Baker report did not stop there. It went on to make the extraordinary claim that the chief executive's preoccupation with global warming may have somehow led to the accident.

Near the end of a much-celebrated career, it seemed that everything was falling apart for BP's chief executive. Even before the Baker report, a row with BP's nonexecutive chairman led to an embarrassing rejection of Browne's request to stay at the helm beyond the firm's mandatory retirement age. After the bad publicity caused by the Texas explosion and the Baker report, Browne decided to fall on his sword and retire even sooner than originally planned in 2007, thus sparing the firm months of negative publicity.

The final indignity came on a night that should have been one of his finest hours. In January 2007, he was due to receive a coveted award from CNBC, the business news network, at its annual gala dinner. The ballroom at the Pierre Hotel was packed with various chief executives, Wall Street players, and the business press; even Alan Greenspan was present. Several executives, including the bosses of United Technologies and Southwest Airlines, got awards, as did Alan Greenspan for his life's work. But the top award for overall excellence (highly coveted, as it was determined by a jury of fellow chief executives) was meant to go to John Browne. BP's American arm, ecstatic about the honor, booked a table and invited journalists along so they could show off the big new prize. When the evening arrived, however, there was no mention of BP or Browne on the program and not even a speaking slot for its American vice chairman, who had prepared a speech for the occasion. The award had been removed completely from the program, with no advance notice or apology given to the oil company. Adding insult to injury, grumbled one senior BP executive present, was the fact that "they've even stuck us here in the worst table at the back of the ballroom."

Did John Browne deserve this ignominious end? He can be accused of overpromising on the green front, to be sure, and it is only right that the boss pay a price when workers' lives are lost on his watch, but the swift demolition of his once-gilded reputation is probably unfair. Look beyond the troubles of the last few months of his career, which ended under the cloud of a court case involving a former romantic partner, and you will see a man who has done more than any other to prepare the petroleum business for the daunting challenges of the future.

Beyond Hype

Put the skeptics' arguments to Lord Browne, and you find that he is surprisingly contrite. Yes, BP will spend more on hydrocarbons than alternative fuels for a long time to come. Therefore, he agrees, BP will remain chiefly an oil and gas company for a long time. But there is nothing evil or hypocritical in that, he argues. He points out that compared to the trillion-dollar industries of oil and gas, green markets are small today. Therefore, even his comparatively stingy investments in them already make him a green giant; these emerging markets cannot absorb more capital quickly. As proof of his firm's commitment to nurture these markets, he points to the $8 billion to $10 billion in fresh money committed to alternative energy over the next years and an unprecedented $500 million pledged in early 2007 to fund cutting-edge research into low-carbon technologies at the University of California at Berkeley.

What is more, the quibbles over renewable energy miss the bigger picture. The reason to think BP does deserve praise arises from Browne's extraordinary courage on global warming, not from all the solar panels it is not putting up. In the run-up to the UN's Kyoto Summit in 1997, the industry (via the GCC) spent a fortune pressing its line that the issue is complete nonsense. That was when Browne broke ranks. In a dramatic speech delivered at Stanford Business School, he openly accepted that climate change is real, that mankind plays a role in it, and that this could have dangerous consequences. More to the point, he broke with the GCC and declared his support for the aims of the Kyoto Summit.

That act broke the energy industry's united front on global warming. As Shell followed in BP's footsteps, it left the American oil industry to follow Exxon's stout defense of the status quo. Since then, Browne committed his firm to reducing its own emissions of greenhouse gases well in advance of any legal requirements. Long before Europe had set up its carbon-trading market, BP launched an internal market for emissions trading among its divisions around the world. In an effort to prod America into action on global warming, Browne even penned an article in *Foreign Affairs* warning of potential tipping points in the climate system that could lead to disaster; he advocated the adoption of a specific numerical target for global emissions of greenhouse gases (at roughly double the level that prevailed during preindustrial times, a target supported by most serious climate scientists) so that industries and governments could best organize their strategic responses.

When Browne broke with the GCC, most oil bosses and environmentalists were even more skeptical than the Sierra Club is about Bill Ford today. They suspected Browne was either a naïve turncoat or a sophisticated fraud. The most cynical argued that Browne was neither green nor a hypocrite. Rather, they argued, he was a brilliant opportunist who saw the standoff over global warming as a business opportunity waiting to be seized. Since the tide was always likely to turn against the oil companies, goes the argument, BP merely took the moral high ground with some early burnishing of its environmental credentials. After all, that is how things have played out: in the years since his break with the GCC, BP has been applauded as a farsighted, ecologically sound firm. Its successful internal emissions-trading system helped inspire the EU to launch the world's first international carbon-trading market. Even some rivals, most notably the European firms, are now following BP's lead.

So was Browne's move on global warming a clever effort to seize yet another "first-mover advantage" over his rivals, the way he stole a march on the industry with his stunning take-overs of Amoco and ARCO? When asked point-blank, the usually unflappable Browne fell silent. On the contrary, he said; BP's top management was concerned that there might be a first-mover *disadvantage*. Oil companies have such a poor public image that the first to profess concern for global warming might quickly suffer a backlash from a cynical public. And fellow oilmen might have ostracized BP, even to the point of damaging its future business ventures.

If that was the advice of his inner circle, then why in the world did Browne set out on the green path? One reason, he explained, was the science: both his researchers and he himself had grown troubled by the mounting, though still inconclusive, evidence of global warming. He was also concerned about the implications: "We simply cannot survive for long if we remain so out of tune with our consumers' perceptions, and the next generation's attitudes." Another factor was a conviction that oil companies "must engage in the debate, and not be shut out as the bad guys. I want us to be . . ." Here he again paused, and continued softly, "I want us to be—dare I say it—progressive."

Such sensitivity distinguishes Browne from many of his rival oil bosses—especially those running American firms. Revealingly, he is willing to acknowledge mistakes made in his firm's overeager advertising campaign. Browne has learned some important lessons about overpromising: "Be very careful to separate aspirations (for society) from actual promised actions." Perhaps most upsetting for advocates of the "triple bottom line" and other sustainable business theories, he is also refreshingly candid in debunking the woolly notion of corporate social responsibility: "Business is about doing business, it's not a surrogate for government or public service."

On balance, it seems that Browne's embrace of the environment was pragmatic but genuine. There will always be naysayers, but BP even managed to win over many environmental leaders through its policy advocacy and technology investments. Environmental Defense, for example, a leading American green group, worked hand in glove with the firm to design BP's internal emissions-trading system for greenhouse gases. Arnold Schwarzenegger, America's greenest governor (thanks to the influence of Robert F. Kennedy, Jr., a relative who happens to be hypergreen), was also won over by Browne's vision for low-carbon energy. The Governator applauded BP's groundbreaking carbon-free power plant, opened in 2006 near Los Angeles. Even the environmentally minded UC Berkeley saw enough of substance in Browne's green vision to join hands with BP in 2007 to develop low-carbon technologies.

This is not to say BP has turned into Greenpeace. Indeed, the biggest battles Browne fought were inside his own firm, much as Bill Ford's greatest challenge was the enemy within Ford. Steve Westwell, the head of BP's low-carbon division, in 2007 addressed an audience at Stanford, where a decade earlier, his boss had broken so sharply with the oil industry. He confessed that internally, it was "not all smooth sailing" running the green division of an oil giant. On public policy issues like carbon regulation, he conceded that "we do stand on both sides of the fence. If we do it well, we can be an honest broker to society in advising on the likely costs of rival policies." But, he conceded with a sigh, "If I get too radical on policy, I would be called back by the rest of the firm, saying, 'Let's be logical here, Steve.'"

That reality check explains why it is wrong to ask companies, be they BP or any other, to behave in ways that ignore their self-interest. Browne was not choosing between saving the planet and saving his own bacon: he was doing both. And only policies that accomplish both aims are likely to be sustainable over the long run. His courage on global warming is best seen as a part of his general eagerness to modernize a business that has long thought it was above change. He imposed a discipline on costs that were inflated by folk memories of the great petro-dollar era two decades ago. In the same way, he demanded that an industry on the defensive should come out fighting, declaring that it is ready to be held responsible for the environmental impacts of its products—as long as government and society lead the charge.

Unlike the formidable Lee Raymond, who could squeeze a dollar out of a nickel but who had his head in the sand on global warming, John Browne had the courage to take the threat posed by climate change seriously. In doing so, he may have done nothing less than start preparing BP for the unthinkable: life after oil. And if indeed he got it right, it will be his rivals in the dinosauric American oil industry who turn green—with envy. History shows the dinosaurs that are able to embrace change, like Toyota, can flourish, but those unable to dance to a new tune tend to go extinct.

Dancing with Dinosaurs

There are few industries with such polar opposites as BP and ExxonMobil. While Browne has transformed his company over the last decade into one poised for a low-carbon future, Lee Raymond spent that time fighting such a future. It is no wonder that Exxon is convinced the Age of Oil will go on and on and on. Exxon's energy forecasts argue that even as far out as 2030, internal-combustion engines will still make up over 95 percent of the world's vehicle fleet and that oil will remain top dog. Lee Raymond loved to declare research into renewable energy "a complete waste of money." He long argued that global warming was an unscientific notion perpetuated by government scientists seeking funding, though his company has tried to soft-pedal such views in its public pronouncements after his retirement.

It is dangerous to argue with a man who created, on one estimate, more than $130 billion of shareholder value during his tenure—and not just because he is likely to chuck you out of his office. However, it is just possible that he was too dismissive of the long-term risk posed to his beloved firm by climate change and geopolitics. If governments move more aggressively to promote alternatives to oil because of those concerns, then even a giant like Exxon could find itself relegated to the garbage pail of history.

After all, IBM and AT&T were both giants dominating their

respective industries a few decades ago. Neither saw the impact that government policies and disruptive innovations would have on their industries. IBM simply did not see the rise of the personal computer, while AT&T was caught out by the deregulation of the telecom industry. These dinosaurs eventually learned to dance, but other once-successful giants, like Digital Equipment and Wang, bit the dust.

If America's oil majors want to ensure a bright future, they would be wise to take the threats to oil's supremacy seriously. BP and Shell have done the most, as each has set up divisions to look into low-carbon energy. Especially given the challenges facing it in replacing reserves, the industry would do well to get serious about alternatives to oil, such as biofuels and hydrogen. Indeed, oil bosses need to expand their notions of what it means to be an energy company if they are to survive and flourish. Sir Mark Moody-Stuart said it best a few years ago, when he was chairman of Shell: "We need to meet our customers' needs for energy, even if that means leaving hydrocarbons behind."

If the dinosaurs don't learn to dance, they may find themselves as irrelevant one day as the buggy-whip manufacturers, who were wiped out by the slow but sure advance of "horseless carriages" a century ago. That is precisely what will happen, argues Jeremy Leggett. He is the head of Solar Century, a British renewable-energy pioneer, and also a leading advocate of the Peak Oil theory. Leggett is convinced that the majors don't have a chance of survival postpetroleum.

"When has any disruptive innovation that has upended an industry ever come from one of the industry's incumbent powers?" he asks. "Radical change can come fast, and it usually comes from outsiders." He thinks even the progressive European oil companies will bite the dust. A visitor to the stylish low-rise headquarters of his firm, on the south bank of London's

Thames River, found him on the roof, solar panels scattered hither and yon. Clean energy is the future, he insisted, pointing accusingly at Shell's headquarters building on the river, "and that monstrosity with the tattered Shell flag on top . . . that is surely the past!"

End of the Oil Age

The most powerful man in the oil realm, Saudi Arabia's oil minister, Ali Naimi, is himself beginning to take note of policies designed to spur alternatives to oil. That was not so as recently as 2000. At a lavish banquet thrown back then at a pleasure palace outside Riyadh, Ali Naimi was relaxing with a favored Saudi prince and a host of foreign oil bosses and diplomats. Dozens of Arabian horses and camels frolicked on the grounds, and tribesmen in traditional attire entertained the gathered kingpins of oil. Even America's secretary of energy at the time, Bill Richardson, took the trouble to attend and pay his respects to the central bankers of oil. When asked about the prospects for hydrogen and a move beyond petroleum, Ali Naimi immediately responded, "Hydrocarbons will remain the fuel of choice for the twenty-first century."

Asked the same question again in 2005 in his ministerial office in Riyadh, he reflected before replying. He had been surprised by the size of the investments in fuel cells now being made by the global car industry. He was concerned about the UN's Kyoto Protocol and other efforts to tackle climate change, which he believed are here to stay and which he was certain will hurt oil. Most revealing, he said that his country is now looking into carbon-sequestration technologies.

Astonishing as it may seem, even Saudi Arabia, sitting atop a virtually endless font of petroleum, is now beginning to hedge its bets. Given this, how on earth can the rear guard of America's oil industry justify its backward-looking, oil-embracing, climate-denying stance? "Oil will still dominate for the next thirty to fifty years," said Ali Naimi with a smile, "as long as there are no meaningful substitutes."

Petroleum will clearly be with us for many years to come, but at last a serious race is under way to find a replacement. If the world suffers a terrorism-inspired oil-price shock or if governments decide to move more forcefully to avoid the worst effects of climate change, then that day could arrive sooner than many imagine. Look beyond the mucky swamp of Washington politics, and you will find that there is reason for optimism. As the next section of the book explains, there are breathtaking forces at work and nimble outsiders from unexpected quarters—ranging from the fuel-cell laboratories in China to the venture-capital incubators of Silicon Valley—already challenging the status quo. The world may yet move Beyond Petroleum.

One key factor that will determine the speed at which this happens will be the future of China, where a repetition of the carbon culture of the West would put an unbearable strain on the planet, as mass motorization doubled the rate of greenhouse-gas emissions coming from autos. But there is good reason to think that will not happen as China leapfrogs to the front of the world stage.

III

MANIFOLD DESTINY

The clean car of the future may come from completely unexpected sources

CHAPTER SEVEN

Crouching Tiger, Leaping Dragon

Asia's rise could save, rather than destroy, the planet

Two hundred years ago, Emperor Napoleon forecast that when China woke up, the world would tremble. The Middle Kingdom had a generation earlier slipped into a mood of isolation that saw it fall from being (with late-eighteenth-century India) the world's biggest economy, as the Industrial Revolution swept Europe and North America to wealth and world dominance. Now, the sleeping dragon is at last waking up and stretching itself—and the world is indeed beginning to tremble.

Thanks to the extraordinary economic growth of recent years, China is the world's fourth-largest economy by conventional measures; if measured in local purchasing power, it is already second only to America. By the middle of this century, its economy will be as big as America's at market exchange rates. It is already the third-biggest car market, after America and Japan. The first Chinese car exports are beginning to reach Europe and the United States. In the ten years from 1995, Chinese car production went from 320,000 a year to 2.6 million. There are already twice as many mobile phones in China as in the United States, and within ten years, the Chinese will buy more cars than Americans. Ten years ago, 94 percent of the goods stocked on Wal-Mart shelves were made in America; today, four-fifths of the retail giant's products come from China.

Some compare Asia's rise to the recovery of Europe and Japan from World War II. But that was puny compared with what is happening in China and India. China has been growing at around 10 percent a year since it opened its economy to the world in 1978. It has doubled its wealth in the past ten years alone. America took almost two generations, from about 1870, to emerge as the world's leading economic and political power by 1918. China is doing everything faster, leapfrogging its way to the top. The example was set in the 1980s and 1990s by four other nations in the region, the so-called Asian Tigers—Singapore, Thailand, South Korea, and Taiwan. They were the first nations to press the fast-forward button for economic development, to go from peasants and paddy fields to cars and condos in one generation. The children of poor subsistence farmers grew up to enjoy life in rich, sophisticated countries. But those tigers are tiny compared with the most populous country in the world. When the biggest nation, home to 1.3 billion people, does the same, the world will indeed tremble. When its steps are being dogged by what will soon be an even more populous country, India, a planetary shift is afoot.

Some economists and environmentalists think the emergence of China and India, in the way that it is having an impact across the whole world, is more akin to the rise of the Roman Empire or the discovery of the New World than to the economic boom seen in postwar Japan and Europe. The Roman Empire laid the foundations for the development of modern Europe, although it took a thousand years for it to emerge via the Renaissance. The effect of the discovery of the New World was altogether more rapid. Shorn of its colonial shackles, an independent United

States took only 100 years to emerge as the economic challenger to the British Empire and a further 40 years to establish itself as the world leader, stepping in to settle World War I. For about 150 years, the British navy had patrolled the oceans and ensured the sun never set on the British Empire. Since the 1920s, it has ceded that global imperial role gradually to the U.S. Fifth Fleet.

It is not fanciful to imagine similar dramatic changes under way now. Between them, China and India account for over a third of the world's population. As they develop, their impact will surely be planetary. Some of that impact will be undoubtedly positive, of course: countless unfortunates in the developing world will be lifted out of poverty, and the poorest in the developed world will enjoy more affordable imported consumer goods and outsourced services even as America and Europe find booming new markets for their higher-value exports. It may well turn out that growing global trade increases everyone's prosperity, rather as a rising tide lifts all boats.

However, there could be a dark side to this coming long boom. A gloomy chorus now argues that the rise of Asia will consume so many commodities and spew out so many pollutants that it will inevitably lead to resource wars and render our planet unlivable. Are they right?

How Many Planets?

"How many planets would it take if everybody in Asia and the Pacific consumed like an average American or European?" That is the question posed by Emil Salim, a former Indonesian minister and development expert, in the introduction to the "Living Planet Report" produced by the World Wildlife Fund (WWF) in 2005. Fittingly, that year's report by the green group, famous for its panda bear mascot, focused on the likely planetary impacts of China's rise. Though Salim did not give credit where it was due, his provocative question was a modern spin on a question posed by Mahatma Gandhi sixty years ago about how many planets it would take if India pursued Britain's resource-intensive path of industrialization.

The argument put forward by the WWF is certainly eye-catching. "As a planet, we are living beyond our ecological means," declares the report. The approach used by the environmental experts can be seen as a sort of balance sheet, or profit and loss statement, for Corporation Earth. The experts first calculate global biocapacity (which is the ability of the planet's ecosystems to absorb wastes like pollution and to generate resources like food) and then deduct mankind's demands on pasture, forests, fisheries, cropland, and so on. According to WWF, "Overshoot is no longer a hypothesis but a reality . . . humanity's annual demand for resources is now exceeding the Earth's regenerative capacity by more than 20 percent."

In other words, they claim it already takes more than one planet to support modern life—and if current trends continue in Asia, argue such experts, we will soon require two planets' worth of resources. (By "one planet," what is meant is the amount of resources that the green group thinks can be regenerated by the earth each year; on this argument, we can run a "deficit" only for a short while.) More than half the world's population lives in Asia, argues WWF, and its use of global ecological capacity has shot up from 15 percent in 1961 to 40 percent in 2001. The result is that Asia requires one and a half times its own land and sea space to support its resource demands—as anybody familiar with the meteoric rise in the prices of oil, steel, and other commodities (which were sucked in by the booming Chinese economy from Africa, Latin America, and elsewhere) during the first half of this decade knows all too well.

The alarming thing about this forecast is the growth that is yet to come. Goldman Sachs looked at four developing giants—Brazil, Russia, India, and China (BRIC)—and drew these conclusions: "The results are startling. If things go right, in less than 40 years, the BRICs' economies together could be larger than the G6 (today's six biggest economies) in US dollar terms . . . only the US and Japan may be among the six largest economies in US dollar terms in 2050."

If—and this is a big if—the BRICs grow that wealthy by following the resource-intensive path laid out by Europe and especially by America, as Salim and Gandhi feared, then an ecological nightmare is in the cards. If indeed the pessimists are right, then the *Observer*, a British newspaper, was right to scream in a recent headline: "Wanted: New Earth by 2050."

Demography Is Not Destiny

Just consider the scale of the potential problem—for instance, the effect on global warming of 750 million more cars in India and China, belching carbon dioxide in both countries by the middle of this century, more than three times today's fleet of American gas-guzzlers. It took over a century after the invention of the automobile to approach a billion-car world, but over just the next three decades, these two giants could take us to a two-billion-car world.

Between them, China and India already account for 17 percent of world energy consumption, with China's oil imports surpassed only by America's and Japan's. China, which used to produce enough oil for its own use, has become the second-biggest consumer after America, whose history it is imitating in this respect, as in many others. By the mid-1930s, America had already seen itself beginning to outstrip its domestic oil supplies and had started to worry about the strategic implications of depending thereafter on imports from oil fields being discovered in the Middle East. China has started scouring Africa for supplies of oil and has tried to buy one American oil company, Unocal, and is courting Russia for its oil and gas.

Currently, China uses only one-fifteenth the amount of oil per head as the United States. India uses half that amount, but both have doubled their rate of consumption since the early 1990s. If both were to increase oil use to about the same level as Japan (which is only about half the American rate per person), argues the World Watch environmental think tank, it would require another 100 million barrels of oil per day on top of the 85 million the world consumed in 2005. If OPEC doesn't produce such astounding amounts of new oil on demand, goes this gloomy argument, then economic depressions and resource wars might break out.

That is a frightening thought, and the eco-pessimists make a strong argument, but that does not necessarily mean they are right. Demographics need not be destiny, as history has shown. Today's Cassandras are making much the same mistake the Club of Rome experts back in the 1970s did in forecasting global shortages of food and commodities; that group of prominent and well-intentioned thinkers even claimed three decades ago that oil would run out. In fact, history shows that Malthusian prognostication was wrong: the world did not run out of food, oil, or any other resource. The flaw back then, as now, is that such forecasts are often just straight extrapolations that do not take into account any positive feedbacks. In the real world, scarcity and crisis breed substitution and innovation. Development is a dynamic dance, and straight-line trends rarely go on forever in one direction.

The current forecasts about oil running out or resource wars do of course contain a grain of truth but still fall wide of the mark. That is because they ignore inevitable market responses, such as the ability of the oil industry to extract oil from myriad untapped reserves and the pressure to develop alternative energy caused by a tight oil supply and high prices. And even if China buys oil assets overseas—as it tried to do with America's Unocal in 2005, only to find itself blocked by a xenophobic backlash in Washington, D.C.—that is actually good news for America, since access to such "equity oil" means China will then be buying less from the world oil markets, easing prices and leaving more for everyone else.

In fact, the market provides the best reason to think resource wars are unlikely to result from Asia's rise. If demand for oil soars, then prices will inevitably soar too—for every single consumer on earth, given that oil is a fungible global commodity. That will push China and India to look harder at energy efficiency and encourage them to develop homegrown alternatives to imported fossil fuels. The likely result is that China and India will lead the world into new ways to use their coal and their huge potential for renewable energy; both countries are now keenly investing in, for instance, wind power.

When delegates at the Euromoney Renewable Energy Finance Forum held in late 2006 in London were asked to vote on the leading "renewable city or region" in the world, they did not pick übergreen Berlin, Germany, or crunchy-granola Berkeley, California. The gathered renewables-finance types gave the award to the southern Indian state of Tamil Nadu, one of the pillars of that country's software and information-technology leapfrog. Tamil Nadu generates over 2,000 megawatts of electricity from renewable sources, over a fifth of its total grid capacity and about half of India's total renewable output. The Indian state squeaked out a victory just ahead of China's Dongtan eco-city being designed in the Yangtze Delta.

K.K. Chan, managing director of the renewables division of China Light and Power, said at the time, "Like in many other industries China presents a huge opportunity in terms of energy development. China wants to jump-start its energy industry without making the same mistakes as other developing countries."

Chan argued that government policies in the shape of mandatory green targets for the power sector (15 percent of the country's energy needs from renewables by 2020) were the key: "The companies have to buy the renewable power because the Government owns the utilities and the Government sets the targets. It's simple! The law recognizes the fact that renewable energy and meeting these targets is expensive, so the Government has agreed to spread the cost and to effectively subsidize. . . . Effectively, the whole country will share the cost. The legislation spans twenty years so it's a twenty-year promise." That is green, but there is no guarantee that all those new domestic energy solutions will necessarily be low-carbon: windmills certainly are, but coal can be used in either climate-friendly ways (with the sequestration of the carbon content) or climate-unfriendly ways (by just burning it in the old-fashioned way or by making liquid fuels out of it in dirty ways).

That is why Asia's biggest impact on the world in terms of resources and the environment is likely to be its carbon footprint. Drill down deep into that WWF "Living Planet Report," and you discover a misleading analytical trick that confirms the fact that carbon is the real problem—and that most other environmental problems are actually under control. The main weakness of the WWF approach is that its assumptions about energy are questionable. The group rightly worries about the greenhouse gas emitted by fossil fuels, but misleads readers when it defines the carbon footprint as the area of forest required to absorb emissions of carbon dioxide (leaving out that absorbed by the oceans).

The world's energy footprint shot up from 2.5 billion hectares in 1961 to 6.7 billion in 1999, by far the fastest-growing factor making our civilization look unsustainable—and revealingly, energy completely swamps all the other factors. Take out energy from the footprint analysis, and you find that suddenly most of the other problems environmentalists complain about with regard to China and America—excessive consumerism, McMansions, the disposable society, aquifer depletion, deforestation, air pollution, and so on—suddenly become much more manageable with one planet's worth of resources. Surprising as it must seem, the ecological footprint per person (leaving out energy) has actually been shrinking over the past few decades. So if Asia increases its energy use by embracing low-carbon energy technologies and leads the rest of the world off fossil-fuel combustion, then even the WWF would have to agree that the world's future suddenly looks much more sustainable.

The Key Is Carbon

Once one leaves eco-utopian ideology behind, it becomes clear that the central challenge posed by the rise of the dragon is this: can Asia really decarbonize, or is it doomed to follow America's paradigm of carbon-intensive growth? As these countries continue to grow at their recent hectic pace, the concern is the greenhouse gases they will spew into the air just as the world is gripped by fears of accelerating global warming. As David Hawkins of the Natural Resources Defense Council points out, unless the "business as usual" approach is altered, China will build dozens of giant coal-burning power plants over the next two decades that will prove a climate nightmare—unless, that is, they capture the carbon using capture and sequestration tech-

nologies that are now available. Unlike some environmentalists who take an unrealistic "renewables or nothing" position,
Hawkins sensibly acknowledges the cold reality that China,
India, and South Africa are poor countries endowed with vast
quantities of cheap coal and therefore will surely use it before
they turn to expensive or imported energy.

Despite the windmills now going up in Dongtan and elsewhere, nearly three-quarters of China's energy still comes
from coal: it extracts 2.2 billion tons a year, twice America's
production, and still has huge reserves. The coal problem is
bad enough, but at least there is some hope of tackling it: low-
carbon sources of electricity like wind and nuclear power are
well established, and the number of power plants in the country is relatively small. If the government wanted to crack down
on carbon emissions, it could squeeze the country's key utility
bosses to clean up their emissions or switch to low-carbon technologies, as Hawkins suggests.

Cleaning up carbon from cars, in contrast, could be a nightmare. Unlike coal, oil has no ready substitute today. Moreover,
each car empowers and emboldens its owner in ways that no
other invention in human history ever has. Even in dictatorial
China, the government does not dare tell the aspiring middle
classes about to buy their first cars that they must ride bicycles
or crowd onto buses. And even if China's leaders tried, they
would probably fail to kill the car.

The Japan International Transport Institute (JITI), a think
tank, has done a detailed forecast of future growth in Asia's
automobile market. After studying trends in Japan, South Korea, and other recently developed countries, JITI argues that
three factors determine when a country crosses a "motorization
inflection point": first, car ownership grows by 10 percent a
year or more; second, the ratio of cars per head rises to fifteen

to twenty per thousand; and third, the cost of a low-end car approaches the average household's annual income. The first two conditions have clearly been met in China, and the third will soon happen. Indeed, if you look at the richer coastal parts of China, where incomes are far higher than the national average, you find that car sales are exploding. But here comes the carbon kicker: JITI experts are convinced that the greenhouse gases spewed out by China's vehicles will double by 2030, even if the government takes the politically risky step of doubling gasoline prices.

Driven by cars and coal, China will soon overtake the United States as the world's biggest emitter of greenhouse gases. Such a vast carbon shadow would stretch over the entire twenty-first century (since carbon dioxide stays in the atmosphere for a century) and would render the carbon cuts called for by the UN's Kyoto Protocol irrelevant. Unless China can decouple its greenhouse-gas emissions from its growth, an awful prospect faces a warming world. Unless it succeeds in its efforts to decarbonize its economy, and especially its cars, the world has no chance of reining in global warming.

The good news the eco-pessimists will never tell you is that there is, in fact, hope on that front. The risks are real, to be sure, but so too are the opportunities for a clean-energy revolution led by Asia's giants. China's economy is on a steep curve, but so too is its determination and ability to do something about air pollution and greenhouse gases by leapfrogging dirty processes and going straight to clean technology in transportation and other areas. José Goldemberg, a farsighted Brazilian energy expert and the godfather of that country's ethanol program, was among the first voices (and for a long time a lonely one) to argue that developing countries must not blindly follow America's example on energy:

Developing countries have a fundamental choice: they can mimic the industrialized nations, and go through an economic development phase that is dirty, wasteful, and creates an enormous legacy of environmental pollution; or they can leapfrog over some of the steps originally followed by industrialized countries, and incorporate currently-available modern and efficient technologies into their development process.

The world is waiting to see if China and India can indeed leapfrog to a cleaner world, perhaps teaching America and Europe something along the way. China needs to act quickly, if only to clean up the soot and smog that darken the skies of its cities. At the same time, it ought to be moving to deal with carbon emissions, rather than imitate the American and European example of cleaning up the air, then sitting back for a whole generation before starting to do something about carbon.

Will such a leap really happen? At first blush, it appears that China is merely following America's lead in embracing the motor car. Look harder, though, and even in the midst of China's petroleum-fueled prosperity, you can already see the beginnings of a greener, cleaner energy economy. And even eco-skeptics will appreciate the most powerful force behind the Chinese push, a force more politically resonant than distant fears about climate change: energy security. The hard men of the national-security apparatus in China (as in America) are growing increasingly nervous about their country's dependence on oil imports. For its entire history until about a decade ago, China was energy independent, even exporting oil; but its economic surge of the 1990s turned it rapidly into the world's second-biggest petroleum user, after America. China's military planners know that their country will grow only ever more dependent on barrels from the

Persian Gulf, home to most of the world's remaining oil—and that they do not have a "blue water" navy that can prevent the U.S. navy from cutting off that vital lifeline in case of a conflict (say, over Taiwan). That security imperative as much as green concerns explains why the Chinese leadership is acting on energy efficiency and alternatives to oil. Ideas floated by midlevel environmental bureaucrats for stiff fuel-economy standards for cars went nowhere for years until oil imports soared; that led the security experts to see that lower fuel consumption would actually make China stronger. And the wake-up call on oil, for the world as much as for China, came with the rise in oil imports seen in 2004.

Oiling the Wheels

America and Europe had grown preoccupied with energy security as they watched the oil price rise from $10 a barrel in 1999 to $50 a barrel and more early this decade, but the Asian giants were still relaxed. By 2005, however, it was the turn of China in particular to be concerned about oil supply, given that its oil consumption skyrocketed by an astonishing 16 percent in 2004. China and India are widely held to have been responsible for about half the increase in global oil consumption that helped drive up prices after 2001. That does not mean the new giants are soon going to suck the world dry: even in these two years, it was noticeable that oil consumption fell in other big consuming countries, largely because they were becoming more efficient at making their economies grow with relatively less oil and energy. And even in China, the eye-popping surge in oil consumption seen in 2004 dropped to a manageable few percentage points' increase by 2005.

The real problem was that the unexpectedly strong rise in Chinese and Indian demand coincided with a period of tight oil

supply. The global spare capacity of oil had shrunk to dangerous levels by 2004. For a number of reasons, such as the leveling off of North Sea and Alaskan production as fields reached
maturity and the underinvestment by OPEC in new capacity,
the cushion of spare capacity had shrunk. That drove prices to
high levels for a longer period than had been seen since the aftermath of the first oil shock, back in the 1970s. It is therefore
wrong to blame China and India for the painful surge in oil
prices seen earlier this decade.

Having said that, it is nevertheless true that the sheer growth in
oil consumption by the two new giant economies is bound to put
upward pressure on prices over time as non-OPEC oil production
begins to slow and even peak in coming years. This, combined
with the erosion of OPEC spare production capacity and underinvestment in existing oil fields and refinery capacity, sets the scene
for extremely volatile and possibly painful energy prices in the
future—as long, that is, as the world remains addicted to oil.

Looking at this development more closely, economists at the
World Bank have modeled various scenarios for the growth of
the economies of China and India and its effect on oil prices.
The bank expects the two countries to roughly double their
share of global oil consumption, from 16 percent, by 2050. By
midcentury, other things being equal, the bank calculates the
price of oil could rise to $133 a barrel (in 2001 prices), com-
pared with $25 at the start of this century. This fivefold increase
is not as outlandish or shocking as it sounds. First, it would
merely be a doubling of the prices that prevailed in the peak
period of 2004 to 2006. Second, we have seen all this before,
even before the 2004 to 2006 oil spikes. The price of Arab light
crude oil rose from $7 a barrel (2005 prices) in 1970 to $40 in
2003—a sixfold increase in little more than thirty years.

These calculations bolster the argument that thanks to the

power of price signals and the dynamism of energy markets, the real trouble arising from Asia's oil-guzzling is likely to be carbon, not oil-price shocks or resource wars. And even that global-warming nightmare might be avoided if China's motorization leaps ahead to clean-car technologies.

The Next Detroit

On the outskirts of Shanghai lies the Jiading District, until 2003 a semirural, mosquito-infested wasteland about 20 miles from the city center. Today it is China's motor city. In 2004, its futuristic steel and concrete structure, containing the country's first motor racetrack, hosted China's first Formula 1 race. The racetrack and a motor museum are intended as a showcase for automobile development. The government has invested over $1 million to develop this auto city, home to a joint venture between Shanghai Automotive Industry Corporation and Volkswagen (the Western car company that vies with GM for leadership in the Chinese market). The idea is to create a center for automotive research and development, where cars and parts are made and where leisure revolves around auto-related activities. What took Detroit one hundred years, say local officials, they are trying to do in just a few years.

Just as the automobile was the vehicle of the United States' economic growth from about 1915 onward, so too has China turned to mass motorization. It did so consciously, seeing the development of its own auto industry as a useful tool for wider economic development and a platform for exporting to the rest of the world.

With the car establishing itself as a pillar of economic development, there is a dramatic takeoff in car ownership, a frenzy of road building, and a booming economy. Between 2000 and 2004, the number of cars sold in China went from two million to five million. China had, by 2005, 21,000 miles of interstate-style highways, double the 2000 figure and second only to the United States. It has over 1 million miles of roads, the third-largest network in the world, half of them built in the past fifteen years. It is a rerun of America's Roaring Twenties and booming 1950s, but all compressed into a few years. Under the Chinese system, the government controls the banks, so it can make this happen by funneling investment into the industry, the cost of capital be damned. Foreign car companies such as GM and Volkswagen—but now including all the big manufacturers—are welcome as long as they form partnerships with local companies and do what they are told. Car demand took off in 2002, growing nearly 31 percent in one year and 45 percent the following year.

The trigger was rising personal incomes. Economists had long speculated that once the average income per head rose to $1,000 (equal to $6,000 in terms of what you get for your money in China), car ownership would take off. A national average income of $1,000 means there are sufficient numbers earning $3,000 to create a demand for cars. In fact, demand grew so strongly that it had to be reined in. The government had to make it more difficult to get credit for purchases in 2004, to avoid inflation. Demand growth for cars settled down into low double digits, and prices started to fall as import tariffs came down, as they had to after China joined the World Trade Organization.

Car ownership is still just around 10 to 20 cars per 1,000 people, compared with a global average of 120 per 1,000 and more than 600 per 1,000 in the United States. There will soon be around ten million cars in China—about half the number in the United States at the time of the Wall Street crash, which sent car sales tumbling by 75 percent. But the parallels with America go deeper than simple motorization. Many of the same forces of motorization are reshaping China's urban landscape.

Until the mid-1990s, large cities were full of people who worked for state-owned factories or other government organizations, living virtually rent-free in cramped apartments owned by their employers, close to their place of work. A bicycle or bus ride was sufficient to get them to and from work. In recent years, however, many such employers have relocated to the suburbs to reduce urban air pollution, and housing has been privatized. The government allowed banks to lend to consumers newly hungry for cars and was happy for developers to do to inner-city areas what Robert Moses did to the Bronx in the 1950s: the bulldozers moved in, and the people moved out. Many people found themselves with long, slow bus rides to their jobs just as they found they were in a position to buy a car for commuting. Beijing has twelve million residents and two million cars. Despite a new light-rail system, public transportation struggles to meet demand, encouraging more people to use cars.

But here is where China and America seem to be diverging. Whereas Big Oil and Detroit combined to drive out public transportation in some cities (for instance, by buying up trolley-car companies and then closing them in Los Angeles), China is now trying hard to develop public transportation as well. Nor are autos the only way the Chinese are getting around under their own steam. Although ordinary bicycles have been banned from many big roads, electric bicycles are coming in force: sales topped ten million in 2005, much greater than the number of cars sold. So successful have they been that carmakers have lobbied city officials in Beijing and Shanghai to keep electric bicycles off some streets to make room for automobiles. Although

not as green as human-powered cycles, these bikes have up to
twenty times the energy efficiency of the smallest of cars.

The Cure from Curitiba

Clean cars are an inevitable part of China's future, and
they can be consistent with a sustainable future—but only if
integrated into a broader transportation policy that provides
consumers with genuine choices of public and private transportation. Happily, after dashing for an American-style, au-
tomobile-fired path of prosperity for some years, the Chinese
leadership is now clearly acknowledging that cars and cars
alone are just not sustainable for a country of 1.3 billion people
without destroying resources needed for other uses.

The ministry of construction has declared that building public transportation must now be a national priority, and it is
pushing for the adoption of bus rapid transit (BRT), a system
first developed in the southern Brazilian city of Curitiba and
since introduced in Bogotá, Colombia, and in Mexico City.
BRT combines the speed of a subway with the construction
costs of a bus. Buses have dedicated lanes, and their drivers can
change traffic lights to green as they approach. There are raised
platforms for passengers getting on and off. Machines allow
them to prepay to speed up boarding. Since first tried in the city
of Kunming, the capital of the southwestern province of Yun-
nan, car traffic there has fallen by 20 percent, and buses have
doubled their share of transport, to 13 percent. During rush
hour, there are five times as many bus passengers as before, and
bus speeds have picked up, to 10 miles an hour from barely half
that. Beijing and Chongqing are planning to follow Kunming's
example.

At a government level, however, the most striking evidence
of a desire to move to cleaner, greener development is the up-

dated Energy Conservation Law that came into effect at the beginning of 2006. This aims to cap energy consumption in 2020 at double the rate of the year 2000, even as the economy quadruples. China is second only to Russia in the energy-inefficiency league—that is, the amount of energy it needs to produce an extra dollar of output. But it has made much progress since 1980, cutting the input of energy per unit of output by some 60 percent.

The country has been passing various environmental-protection laws since modern development started in 1978, but they have been largely ineffectual. Of the twenty cities worldwide with the most polluted air, sixteen are in China. About two hundred Chinese cities are estimated to fall short of the World Health Organization's (WHO) standards for particulate matter (tiny soot particles) that causes respiratory diseases. The high levels of sulfur dioxide in the air also give it the world's worst incidence of acid rain, which now inflicts excessive acidification on one-third of its croplands. The damage of all this to farms and forests is estimated to be around $13 billion a year. The WHO estimates that a quarter of a million Chinese die each year because of poor air quality in cities. The danger is that over the next few decades, the damage caused by the emissions from dirty coal-burning power stations will be augmented by a growing brew of automotive emissions. But at the end of 2005, the State Environmental Protection Administration announced plans to spend around $160 billion on environmental protection between 2006 and 2010, twice the rate of spending in the previous five years. The latest government five-year plan acknowledges the need for a "harmonious society," which involves improvements in citizens' welfare and the environment. Thanks to tough new emissions laws, the new Buicks rolling off the assembly lines in China spew out barely 2 percent of the tailpipe pollutants of comparable models from 1970.

3-D Vision: Develop-Despoil-Detox

Many observers are convinced that China is engaged in a new sort of Cultural Revolution, turning its back on the twentieth-century 3-D development model of develop-despoil-detox in favor of getting it clean and green from the outset. New Energy Finance is a London business specializing in developments in Chinese energy. Its founder, Michael Liebreich, sees progress across a broad front. For instance, he and his team of researchers think that by 2020, China will be getting up to a fifth of its energy from renewables, up from the official 15 percent target and nearly three times the share in 2006, which at 7.7 percent already made China the world leader.

In the transportation sector, biofuels will meet nearly 10 percent of China's needs. China by 2006 was already the world's third-largest producer of bioethanol. Research labs are looking into cellulosic ethanol, an advanced and efficient form of this biofuel. China has invested some $25 billion to use the Fischer-Tropsch process (which was developed not by America but another coal-rich developing giant, South Africa) for converting coal into a liquid fuel that can power cars and buses. The Clean Development Mechanism, set up under the Kyoto Protocol, allows China to provide emissions credits to industrialized countries in the West in return for money that must be spent on cleaning up polluting Chinese factories and power stations. China has emerged as the biggest source of such credits, ensuring a growing flow of money to clean up its own act.

Even more encouraging is evidence of China's emergence as a powerhouse of new intellectual property. The *Financial Times* reported in late 2006 that China had overtaken Germany in the global ranking for patent applications to become the fourth-largest source of filings; Japan and the United States came in at the top of the list. It seems China saw an astonishing sevenfold increase in patent filings over the past decade, as its economy soared to new heights. The country's policy of welcoming foreign investment and insisting on technology transfer is paying off, as about half of the patents filed from China belonged to foreign companies operating in the country.

The country has clearly become a global hotbed of innovation. The *FT*'s former China correspondent, James Kynge, described the tantalizing prospects raised by this phenomenon in his book *China Shakes the World*:

It is beyond doubt that a degree of cross-pollination between the foreign laboratories and their Chinese counterparts will occur. Thus, for reasons that have little to do with indigenous research, China is leaping up the technology ladder. As every year passes, the conventional wisdom that the country will languish for years in the domain of mid-technology looks more and more like wishful thinking.

And although domestic R&D spending is limited, some national science programs have been strikingly successful. The most famous is the space program, following astronaut Yang Liwei's successful space orbit in 2004. In the construction of powerful supercomputers to help in complicated scientific research, progress has also been impressive. A decade ago, China did not have a single supercomputer ranked among the top five hundred in the world. But by the end of 2003 it had nine, and its fastest, DeepComp 6800, built by Lenovo, was ranked fourteenth.

Its biotechnology is also world class in some areas.

Its scientists are developing a safe "pebble bed" technol-

ogy for nuclear power stations and a "clean coal" system that may allow China to derive energy with vastly reduced carbon emissions. It is, in fact, difficult to think of an area of technology in which China does not have credible ambitions to lead the world.

This evidence suggests that China is serious about avoiding the pitfalls of dirty energy and that it is not being unrealistic in striving for green growth.

Sometimes, clean technologies are more expensive than dirty ones, and sometimes domestic fuels (like coal) are cheaper and easier to use in old-fashioned ways than are newfangled gadgets like solar panels. Nevertheless, the developing giants—led by China but followed by India, South Africa, Brazil, and others—are showing clear signs of leapfrogging. In some cases, as with China and fuel cells and Brazil and ethanol, they are not leaping to catch up with the rich world but instead are leaping way ahead of the current energy infrastructure found in America and Europe.

The Twenty-First-Century Detroit

Every year Michelin, the French tire company, runs a competition called the Bibendum Challenge (named for the obese Michelin Man, the company symbol) for alternative vehicles. It is rapidly becoming a sort of Olympic Games for eco-friendly cars, a demonstration of the latest technology for sustainable transportation. In 2004, it was held in Shanghai, marking the Jiading District's emergence as a motor-industry center. Cars are rated according to their emissions, including carbon dioxide; their fuel economy; noise; braking; and general handling. One Chinese hybrid picked up an award that year, but two years later, at a Bibendum event held in Paris, China walked off with no fewer than four gold medals. This was all the more impressive since China only started thinking about alternative vehicles in 2001.

That was when the national high-tech research and development program known as Project 863 (it was founded in March 1986) decided to set up a program to work on electric vehicles. It headhunted Gang Wan, an engineer who had spent the 1990s working on research for Audi in Germany; he has several patents under his belt for the work he did there on car manufacturing. Back in his homeland, he was charged with developing a strategy for the technical development of the country's auto industry. He decided it was pointless trying to compete with a hundred years of Western know-how on internal-combustion engines. Instead Wan, who also serves as the president of Tongji University in Shanghai, has decided to concentrate on various forms of alternative vehicles—electric, hybrid gasoline-diesel electrics, compressed natural gas (CNG), and hydrogen-fuel-cell–electric.

By the time of the Olympics in 2008 (already dubbed the "Green Games" by Beijing), the plan was to have more than one thousand clean buses and cars on Beijing's roads, with one hundred of them powered by hydrogen fuel cells. In Shanghai, Wan was working on a demonstration program to have one thousand fuel-cell buses and taxis, fueled from twenty hydrogen filling stations. Longer term, the plan is to have the mass production of fuel-cell cars up and running by 2020. That would probably put China ahead of the rest of the world.

The idea of China embracing hydrogen is the most dramatic leapfrogging scenario, but it is not at all unthinkable. Stan Ovshinsky, the octogenarian clean-energy champion and founder of America's Energy Conversion Devices (ECD), even argues that China's hydrogen leap does not have to wait for fuel

cells, as conventional internal-combustion engines can be modified easily to run on hydrogen fuel. The result would not be the
perfect zero-emission outcome made possible by fuel cells, but
it would nevertheless be a very clean stepping-stone that would
make it much easier to roll out hydrogen-fuel infrastructure.
To prove the point, ECD has converted a Toyota Prius hybrid-
electric car to run on hydrogen instead of gasoline. BMW has
gone one better, unveiling a new sedan whose engine can run
on either hydrogen or gasoline—changing fuels at the touch of
a button.

China is particularly well placed to leap to hydrogen because
of the diversity of the energy sources it could use to make that
energy carrier. The hydrogen could be derived cleanly, and with
virtually no carbon-dioxide emissions, from renewables or nuclear energy. Coal could prove a squeaky-clean source for that
hydrogen, if used with techniques for sequestering the carbon
from the coal (thus yielding hydrogen, the other constituent of
any hydrocarbon).

The International Energy Agency, a sober intergovernmental outfit not known for jumping on green bandwagons or
techno-fads, conducted an eye-opening study that suggests
China's vast hydroelectric potential holds the key. The agency
projects that the country's big dams (including the controversial Three Gorges Dam, the largest in the world, with a power
output greater than a dozen nuclear power plants) will run at
a load factor of only 40 percent or so due to the lack of energy-
storage capacity, especially in the rainy season. However, if
the load factor is increased to 70 percent, then all that extra
power, which would otherwise just be wasted, can be used
to make hydrogen via simple and cheap electrolysis (which
splits water into hydrogen and oxygen by applying electricity). That would be enough hydrogen to fuel 37 million cars

by 2010 and 56 million cars by 2020—the vast majority of China's fleet.

China has several huge advantages in going down this path to sustainable transportation. It does not already have a big investment in infrastructure for gasoline-powered cars. Yes, there are lots of cars and roads in China—but this is but a small fraction of the country's future transport infrastructure. In contrast, America and Europe are already pretty much fully built out on roads and gasoline stations, so any newfangled alternatives—be that dedicated rapid bus lanes or ethanol refueling stations—would appear more expensive, since the dirty infrastructure they replace would have to be scrapped, potentially at great cost. China is early enough in the development game that it has the genuine and less expensive option of going green. So China would not suffer from what economists call the "stranded assets" problem—the reluctance to switch to something new for fear of wiping out past investment spending.

Second, China is such a vast country, growing so fast, that it presents a very attractive domestic market in its own right for rolling out new technology (unlike, say, Peru). Japan's rise shows the advantage of having a big domestic market, for its innovative companies typically launch imperfect "soft" versions of cutting-edge products (including Toyota's Prius) at home before daring to export them to America or Europe. Significantly, GM, which has invested more than $1 billion in hydrogen-fuel-cell–electric cars, has signed a collaboration agreement with Shanghai to develop a fuel-cell prototype there and to work with the city on how to develop an infrastructure to support fuel-cell cars.

The third advantage China has in this respect is the power of the state. Unlike many developing countries, where most regulations are respected only in their breach, the local and federal authorities have meaningful power to enforce green mandates if they choose to do so. If the national government and the mayor of Shanghai decide that the only way to cure the city's lousy air quality is to ban all but electric or fuel-cell cars from the city, then it will happen. In fact, the city has already taken a step in that direction, banning all filthy two-stroke engines (of the sort found in cheap two-wheelers and three-wheeled auto rickshaws, *tuk-tuks*, and the like) from the downtown area. That has led directly to a boom in investment in cleaner forms of entry-level transport, such as electric bikes, motorbikes with cleaner four-stroke engines, and even fuel-cell–powered two-wheelers.

While China forges ahead with its own plans for clean vehicles, such as the fuel cells being developed in Shanghai, it would also make sense to harness Western expertise. That is why the first joint venture on hydrogen with GM could be a harbinger of the future. As that company was heading into stormy waters in the early years of the century, its boss, Rick Wagoner, once remarked that his hopes for the future were growth in America's population and in China, where GM has long had a leading position. Perhaps GM, with its fuel-cell technology, will be able to save itself in China even as it helps save the world. GM, like most other automakers, assumes that the long-term answer to auto emissions is fuel-cell–electric cars; collectively, the world's automakers have spent billions of dollars on this futuristic technology.

But like a summit hidden behind successive ridges, the goal of the car emitting only water seems to keep receding. One veteran Ford engineer notably told her colleagues in her retirement speech that she had come into the industry decades ago, when fuel cells were seen as the future. Thirty-five years later, she was bowing out, and fuel cells were still the future. One reason that future keeps getting pushed out is the ingenuity that is going into designing better internal-combustion engines and alternative fuels to run them on. To get to the sunlit uplands of hydrogen, the industry has to wade through a thicket of bio-fuels and exotic hydrocarbons. The next chapter will chart a path through them, asking and answering the fundamental question: are they mere palliatives, prolonging oil addiction, rather than the clean revolution that is really needed?

One thing is clear already, however. Shanghai has the op-portunity to become the Detroit of the twenty-first century, with revolutionary products exported to the whole world as other countries move over to electric and fuel-cell–electric cars. In America, the two biggest obstacles to fuel-cell cars have been the lack of an infrastructure to supply hydrogen to gas stations and the burden of historic investment in internal-combustion vehicles. China's huge, untapped market, its lack of obstacles, and its political system are ideally placed to leap over all such obstacles if the government has the vision to seize the opportunity.

A Great Leap Forward

Technology leapfrogging is not as esoteric a concept as it might seem. In fact, it has been one of the most powerful forces for change in the developing world over the past twenty-five years. The first widespread example was mobile phones. In China, India, and most of Africa, the number of cell phones far exceeds the number of landline phones. For decades, only the elite had working phones, and even the middle classes often had to wait for months or even years for the corrupt and inept telecoms monopoly to install a new landline. As the technology for mobile telephony emerged, developing countries embraced it far faster than did the rich world, despite the obvious fact that people are poorer in those countries. Why? Because it was clear that mobile networks are easier, cheaper, and faster to deploy than fixed-line networks.

The leaping did not stop with cell phones. The astonishing march of micropower is testament to that. There are between 1.6 billion and 2 billion people on earth who do not have electricity in their homes, thanks to the failings of the centralized power grid. Again, the cosseted urban elites typically get subsidized electricity, while the poorest of the poor are left in the dark. They are forced to use inefficient and inconvenient solid fuels, be that twigs and timber or agricultural residue or cow dung, in filthy makeshift stoves and lanterns. The resultant indoor particulate pollution kills so many millions that the World Health Organization thinks it is one of the leading preventable causes of death—a human tragedy on par with malnutrition.

The good news is that the poor are leapfrogging in energy too. Micropower—small generators using microwind, solar, or other distributed forms of power generation—is bypassing the power grid altogether. The World Bank is helping finance a project to bring electric lighting to rural Kenya and Ghana, where people do not have grid electricity and rely on sooty, polluting oil lanterns for lighting. By leaping straight to low-energy-consuming light-emitting diode (LED) bulbs rather than Edison's 130-year-old incandescent lightbulbs, villages will get electric light from LED battery units that can be charged by solar power.

China has just developed a small, magnetic levitation wind turbine that, by eliminating friction, can respond to light winds. It produces 20 percent more electricity than conventional small turbines at half the cost. These windmills will be used to power streetlights on rural roads, using only the airflow of passing traffic. A more recent example of technological leapfrogging applied specifically to address environmental concerns is adoption by the Chinese of energy-efficient refrigerators that do not use the chlorofluorocarbon (CFC) gases that damage the ozone layer that protects the earth from excessive solar radiation. Being efficient consumers of electricity, the refrigerators contribute less than older models to global-warming gas emissions. Thanks to the project, which began in 1989 with help from America's Lawrence Berkeley National Laboratory, by the time refrigerators became a mass-consumption item in China in the late 1990s, the eco-friendly models were ready. Three out of four urban Chinese homes now have fridges, and the country produces well over ten million units a year, the biggest fridge industry in the world, which leapt from nothing straight to eco-friendly technology.

Work has recently started on another dramatic example of leapfrogging that suggests China's environmental future need not be as bleak as some suggest. With 300 million people moving from the country to urban areas, China needs to build several dozen towns and cities a year. It has commissioned Arup, a British-based civil engineering group, to build self-contained, self-sustaining eco-cities. Dongtan will be the world's first purpose-built eco-city. It is being constructed in marshland on an island at the mouth of the Yangtze River, near Shanghai.

No buildings will be more than eight floors high. In an echo of Bill Ford's eco-renovation of the historic Rouge assembly plant outside Detroit, Dongtan's roofs will be covered with turf and vegetation that provide natural insulation and also help recycle wastewater. Giant windmills will capture sea breezes and generate power, while fuel-cell–electric buses will carry those who are not using the ample walkways and cycleways to go from building to building. Organic waste will be burned in an incinerator to augment the electric power supply. Each building will have its own smaller windmill and its own photovoltaic solar panels. Obviously, one showcase town by itself will be irrelevant unless that model inspires many, many others—and only time will tell if that happens. Nevertheless, it is fair to say that by embracing the twin principles of ecological sustainability and the microgeneration of electric power, China is at least headed in the right direction to meet the critical ecological challenges of this century.

So China has already demonstrated that it can leapfrog to new, cleaner technology. This has been noticed elsewhere in the world. GE, one of America's most successful and admired companies, started a makeover of its business in 2005 (using the phrase "ecomagination") to emphasize environmental needs across its range of products and services. A big reason is the huge demand for green technology from developing giants like China, countries that GE envisages might constitute more than half its growth in revenues over the next decade or so. A study conducted by Deutsche Bank in 2006 noted that China was set to spend hundreds of billions of dollars on environmental technology over the next twenty years, presenting a huge opportunity for Western firms with expertise to offer.

China needs to keep up its precipitate growth to provide work for its huge and growing population, with its rapidly rising aspirations of wealth. Yet it cannot proceed as before, without switching to a more sustainable model. Huge sums have to be spent on improving water and sewerage systems and on cleaning up local air pollution. It cannot afford the sort of toxic river spills from chemical factories that so shocked the world in 2005. So demand for big environmental investment is probably greater in China than anywhere else in the world. That should make it a magnet market for Western firms with technology to sell.

Hang on a minute, though. It's obviously true that China

and India are booming. They may crave the latest and cleanest technologies, but are they ready to pay Western prices for them? Push GE hard, and it concedes that profit margins in China are, in fact, tight—especially in the energy business. In late 2005, one GE boss bragged that he had just sold no fewer than three hundred superefficient locomotives there. But when pressed about their profitability, he confessed that only the last hundred of that order is likely to make big money, as those will be the ones built with Chinese labor and parts. China wants the latest green technologies, but given its scarce resources (it is, after all, still mostly a poor country), it will pay only rock-bottom prices for them. That splash of reality explains why, despite the leadership's genuine desire for a clean future, China sometimes makes dirty choices rather than clean ones.

There is nothing easy or inevitable about leapfrogging. To see how hard it can be in a hyperdemocratic and corporatist country rather than a green dictatorship, look no further than India. It took two decades of battle and a combination of bottom-up activism and top-down judicial mandates before New Delhi finally cleaned up its air.

A Particular Problem

On paper, India's capital city seemed to be a sustainable-transportation success story. Despite economic growth that sent its population soaring from two million half a century ago to fourteen million today, the city's traffic was not paralyzed like Mexico City's. And unlike Bangkok, whose fractious politics prevented the building of a much-needed subway system for several decades, New Delhi had a good track record on developing alternatives to the private automobile. Indeed, when experts ranked the world's top cities in 2002 in terms of public transportation, it made the top-twenty list.

But as anybody who has actually set foot in the city knows, the city's transportation system was a disgusting menace to health. People took buses, to be sure, but these decrepit diesel death traps were a poor excuse for a public conveyance. Even worse were the three-wheeled auto rickshaws and two-wheelers, which were often powered by dirty two-stroke engines. The problem was made worse by the fact that a misguided system of fuel subsidies and price controls (for kerosene, among other things) encouraged the adulteration of fuels, which resulted in even worse tailpipe emissions. Taken together, the health impacts of the city's foul air cost Delhi somewhere between $100 million and $400 million, according to a World Bank study conducted in 1995.

Despite that bleak outlook, the good news is that India's capital has made dramatic strides in cleaning up air pollution in just the past few years. The solution arose not from some magical techno-fix, though technology did help, but from a complex and organic mix of grassroots environmentalism, a hyperactive press, much political jockeying, and an unusual degree of legal activism. Back in 1985, M. C. Mehta, a pioneering public-interest lawyer, filed a suit demanding that the Supreme Court force New Delhi to enforce air-pollution laws that it was blatantly ignoring.

Years of resistance followed. A crucial factor tipping the scales in favor of change was the dogged campaign of activism and education spearheaded by the Centre for Science and Environment, one of the world's leading activist think thinks. In 1998, the Supreme Court ordered all public vehicles in Delhi to be converted to compressed natural gas (CNG). This unusually specific technology mandate displeased supporters of hydrogen and other fuels and upset experts who argued that markets know better than judges which technologies best meet emissions targets. Nevertheless, argues a thoughtful study done by the Washington think tank Resources for the Future (RFF), CNG was a relatively cheap choice that was undoubtedly cleaner than the diesel and mixed fuels that were on the road. The industry dragged its feet, and the government ignored the order until the Supreme Court (emboldened by public and press opinion) once again stepped in and demanded action in 2002. By the end of that year, all diesel buses and non-CNG two-stroke engines were kicked off the city's streets.

The city's air is now so much cleaner that this model is being rolled out in other parts of India, and cities around the world are studying its implications. The RFF experts draw these sobering conclusions from this less-than-ideal but nevertheless successful experience with pragmatic, partial leapfrogging (perhaps better called hopscotching?):

In the real world, the choice of environmental policy instruments must meet a number of criteria. Policy choices must be politically acceptable to a wide range of stakeholders. They must be supportable by existing institutions, notably the legal system, available human capital and infrastructure, and by the dominant culture, traditions and habits of each country. Tools and goals must be in line with domestic resolve, will and readiness to perform, since the implementation of environmental protection requires so much of so many actors in society. Technology must be available and accessible.

Solutions in the wealthier nations have generally been imperfect compromises built around all these factors. The challenge for environmental policy makers everywhere is how to strike a workable balance between innovation and the capacity of society to absorb social and technological changes.

As Ruth Greenspan Bell of RFF, the lead author of its study and a skeptical analyst of this topic, likes to say, "If you really watch frogs leap, they don't just leap forward . . . they leap sideways, backwards, and every which way." Overambitious attempts at leapfrogs that ignore the cold, hard realities of developing-country life, she warns, could end up the same.

No Free Lunch

As the Delhi example makes clear, getting the future of transportation right in developing countries will not be easy. Leapfrogging is difficult stuff, and there will be backsliding along the way. For one thing, there are plenty of barriers in the way. Developing countries are, by definition, still relatively poor, and governments, companies, and individuals often lack the resources needed to invest just that little bit more up front that is sometimes needed to make a cleaner energy choice (despite the fact that such investments, like solar panels, pay back the initial capital investment quickly, since the fuel is free). Poor countries often (though not always) lag in engineering and other technical skills, so it is unrealistic to expect the latest microturbine or electronics-laden hybrid car to be embraced in, say, rural Haiti. The biggest obstacle, though, is the inability or unwillingness of governments to provide incentives and enforce regulations—call it sticks and carrots—that demand that foreign investors and local firms clean up their acts.

Even China, which now is rich enough and determined enough to try leapfrogging in earnest, did not go green in the early years of automobile production. That is the tough-minded conclusion drawn by Kelly Sims Gallagher in *China Shifts Gears*, a study of the country's experiences in the car sector. After scrutinizing various foreign joint ventures in the country, the Harvard expert argues that "even though cleaner alternatives existed in the United States, relatively dirty automotive technologies were transferred to China in the 1980s and 1990s." That is not to say she pooh-poohs leapfrogging—she does not. On the contrary, she is convinced it is key to the country's future and entirely doable—but that "it does not automatically occur."

She even distinguishes two kinds of potential leapfrogging—catching up with the rich world quickly, as Africa has done with cellular phones, versus leaping way ahead of the rich, as South Korea has done with its supermodern steel sector built from scratch—and even keeps an open mind about the notion that China could pull a Korea-style superleapfrog on fuel cells and hydrogen. But the key to ensuring that past is not prologue, she insists, is intelligent public policy "to shift gears and create the necessary incentives for the foreign and Chinese firms to change their past behavior."

Charles Leadbeater and James Wilsdon of Demos, a British think tank, agree wholeheartedly. They argue in "The Atlas of Ideas," a thoughtful review of Asian innovation, that South Korea's leapfrog thus far is but a small taste of what is to come: "A country that was on its knees at the start of the 1960s, with few natural resources, a limited higher education system and little or no research, has succeeded through a mix of brainpower, hard work, state direction of large companies and U.S. support, to become one of the most technologically adept and best-educated societies in the world." Developing countries wanting to shift gears like this must embrace a comprehensive strategy that starts with technical education, both at home and through sending students abroad. That is something the Asian tigers, and Japan before them, did superbly—and which China and India, with their homegrown excellence in electrochemistry and software, are now emulating.

The policies that must go hand in hand to foster technological advances, argues Gallagher, include raising the price of gasoline and tightening the screws on air pollution and the fuel efficiency of vehicles. Most developing countries fare poorly when it comes to energy pricing: analysis by the International Energy Agency shows that they tend to subsidize fuels at twice the rate of rich countries, supposedly to help the "poor" (though, in reality, that subsidy is siphoned off by fat cats and middle classes). But after the oil-demand shock of 2004, China did start phasing out price controls and other subsidies for gasoline, and other developing countries are following. China also passed tough new fuel-economy laws (which means new cars in that country are more efficient that ones sold in America today), and it now has relatively tough standards on tailpipe emissions from new cars.

The good news is that there is already hard evidence from within China that such policies do make a difference. Shanghai has the most progressive, and probably the most powerful, city government in China, and it has put in place a rich suite of policies designed to control the growth in motorization; in contrast, Beijing has been slower to adopt such measures. The result of Shanghai's "limitations on the number of motor vehicle registrations and specific policies to restrict vehicle ownership," say the experts at JITI, is that there are more cars per thousand people in Beijing despite the fact that Shanghai is a richer city. In other words, public policies do work, even in sprawling, developing cities. Other policies that show promise at taming the beast include congestion charging (Singapore pioneered this concept, but London and other cities now have successful road-pricing programs), car pool lanes, and stiff taxes on inner-city parking.

Of course, merely cleaning up cars and curbing car use makes no sense unless rival forms of transportation are encouraged in their stead. Take a step back from the fray, and it becomes clear that a sustainable-transportation future is possible only if countries pursue holistic leapfrogging that goes beyond just fuels and engine technologies (important though those are). As the success of the BRT system shows, many people—though probably not all—would probably use public transportation more frequently if it were sufficiently convenient and affordable. They might even walk or ride bicycles, and not just because they are too poor to afford a car, if dedicated walkways and bike lanes were installed by cities in a comprehensive way.

World Bank experts agree with Gallagher on the importance of public policies in managing urban transportation in developing countries, arguing in a report that countries need green policies that both "target the technology of individual vehicles and their fuels, and those that are concerned with the management of the urban transport system as a whole. Both approaches are equally important." As New Delhi's struggle with air pollution from belching buses and smoky auto-rickshaws suggested, a mass move to public transportation in and of itself is no guarantee of sustainability if the buses involved are sooty and sulfurous. This is especially true given that no matter how green we go, the car will likely always be with us, so we'd better take the pragmatic approach of cleaning up and controlling cars so we can learn to live with them instead of wishing them away.

That is another lesson that arises from Shanghai's experience. As soon as the ordinary Chinese become wealthy enough to buy a car, they happily abandon public transportation. Shanghai's economic boom has been accompanied by an annual rise of 15 percent in the number of cars in the past few years, which explains the city's miserable traffic and smog. Officials have tried to curb this by introducing an auction system for new car permits but have been taken aback by the demand. The price of new permits shot up past $5,000 per car, and demand was still strong.

In short, public transportation is vitally important, but it will never dislodge the car. For the world's aspiring billions, the car is the ultimate symbol of status and freedom, even if it perpetuates mankind's addiction to oil—for now. That points to more practical ways to tackle petro-dependency: boosting alternative fuels and engines (the subject of chapter 8) and fixing public policies that unfairly favor oil so that there is a level playing field for clean alternatives (the subject of chapter 9). On both counts, another of the BRIC giants points the way forward.

Carmaking in India used to be a matter of old Fiats or Morrises being assembled from kits sent over from Europe. Those big old carmakers are all still there, but the field was long ago taken over by Japanese and South Korean manufacturers. Now India is getting its own homegrown car industry—one that would appeal even to Mahatma Gandhi, because it is focused on domestic production and simplicity. Tata Motors is the second-biggest carmaker in India, behind Maruti (a joint venture of locals and Japan's Suzuki). Quoted on the New York Stock Exchange, Tata Motors is the biggest part of the sprawling Tata conglomerate. Ratan Tata, the chairman, is a nephew of the famous J.R.D. Tata, whose father—a Parsee textile merchant in Bombay—founded India's first industrial business. The Parsees are a Zoroastrian sect, originally from Persia, that settled in Gujarat on the Malabar Coast of western India. One of the Parsees' defining ethnic and religious practices is their form of disposal of bodies after death. They believe that earth, wind, and fire should not be contaminated by the process of disposal, so Parsee bodies are left on the Tower of Silence outside Bombay to be consumed by vultures until only the bones are left. But the population of vultures is shrinking even faster than that of the inbred Parsee clan because of modern pesticides. This is posing a problem for the Parsees today.

The most famous Parsee of modern times was the Tata patriarch. JRD (as he was known) was a colorful figure, equally at home in India, Europe, and America. He founded Air India in the 1930s, only to see it nationalized after the country won independence from the British. JRD's jet-setting lifestyle made him famous in the 1950s: he adored his glamorous French wife and stayed in his French château when he grew tired of the heat of Bombay. He was to India what the glamorous Fiat boss Giovanni Agnelli was to Italy.

His nephew Ratan could not be more different. He may be less flamboyant, but as a businessman, he is mighty impressive, one of the leaders of a new generation of Indian businesspeople who are prepared to take on the globalized world. He stunned the established giants of the steel industry in 2006 by outbidding the competition to take over Corus, the former British Steel and Europe's second-biggest steel company.

Ratan is a rather shy, diffident man in his seventies who looks younger than his years. His only flashy habit is his tendency to pilot the corporate jet. Otherwise, he follows in the Tata family tradition of conducting socially responsible business. Workers in the Tata steel mills in northeastern India are housed in a company town with amenities such as schooling and health care provided and organized by the company. When Ratan became chairman in 1991, Tata was a sprawling empire in which the family interests were down to 5 percent, on average. An aging band of bosses ran three-hundred-odd companies ranging from trucks and steel to hotels and software—very badly. Ratan got rid of them all and found ways for the family's two charitable trusts to regain control. The group is down to only one hundred or so companies, and almost all are profitable. In the late 1990s, Ratan conceived the idea of an all-Indian car company, encouraged by the success of the Tata truck firm in making light trucks.

He called in an expatriate Indian, Kumar Bhattacharya, who was a professor at the prestigious Warwick University in the English Midlands and a former adviser to Margaret Thatcher. They made an odd couple: Kumar fat, jolly, and expansive over a bottle or two of his beloved Côtes du Rhône; Ratan the reserved, do-gooder, American-trained architect whose main leisure activity is walking his dog on the beach. Egged on by Bhattacharya, Tata developed the first Indian-designed car, the Indicar. After a shaky start, it became a big success. Now Ratan is expanding the range of cars, including his revolutionary new small car. The One Lakh car (named after the *lakh*, an Indian unit of measure for the amount of rupees it is supposed to sell at), is due to hit the market at the end of 2008. Its price could possibly rise, but if plans come to fruition, this car will cost less than $3,000.

It is a four-door, five-seat vehicle designed to meet international safety standards with a simple, tiny engine of less than 1 cubic liter capacity, producing a humble 30 horsepower. Americans might scoff that this would not even count as a motorcycle in their country, but such a cheap and cheerful car might well prove attractive to those in poor countries ready to trade up from motorcycles. Combining his sense of social responsibility with his business sense, Tata planned to have it built in a new factory in West Bengal, an area sorely in need of industry and jobs. He saw a huge market in putting the millions of

Indians who ride dangerously overloaded scooters onto four safe wheels.

The One Lakh car could do for rural Indians what Henry Ford's Model T did for early-twentieth-century Americans or André Citroën's 2CV did for French farmers impoverished after World War II. Ratan Tata's original dream was that this car would make budding entrepreneurs of tens of thousands of Indian mechanics in small towns and villages across the subcontinent, as they would assemble the cars from kits stamped out in a Tata factory in West Bengal. Now he accepts this will happen in a later phase. But he is not the only entrepreneur looking to give Indians four wheels rather than two. Venu Srinivasan is the head of TVS Motor Company, a leading maker of scooters, who rescued that family business from near ruin at the beginning of the 1990s. Srinivasan, like Ratan Tata, was educated partly in the United States. The scion of a Brahmin Hindu family from southern India, he studied business administration at Purdue University. Although born into one of India's richest families, he had to help pay his way by selling bibles in North Carolina during one college summer vacation. "It was eighty hours a week of having the door slammed in your face," he recalls. These days, when he is not joining Ratan Tata in the race to produce a relatively eco-friendly tiny car for the Indian masses, he spends his time and some of his considerable fortune on the restoration of the Jain temples of southern India.

Another rival, also a maker of scooters, is the family firm Bajaj Auto. The firm is run by the young Rajiv Bajaj, who has spotted the tack that Ratan Tata is taking and is equally determined to move into the modest, basic, and highly economical car market. Whereas his father, Rahul Bajaj, was a prominent member in the 1990s of the Bombay Club, a group of dinosauric businessmen devoted to holding back the tide of trade and economic liberalization that threatened the established family dynasty of India Limited, Rajiv seems to be going with the progressive tide led by Ratan Tata.

Another two contenders for making cheap people's cars in India to serve the emerging markets all over South and Southeast Asia are South Korea's Hyundai, which has been successful with conventional car models in America and Europe, and Italy's Piaggio scooter company. Piaggio, part of the industrial network of the Agnelli family (the force behind the European Ferrari, Fiat, Maserati, and Alfa Romeo brands), plans to design its basic car in Italy and manufacture it in India. Even the giants of the car industry around the world are beginning to pay attention to these economical, ultra-low-price vehicles. Toyota's boss Katsuaki Watanabe says the time has come to employ revolutionary new techniques to make cheaper cars, even ones involving new, lighter materials. The European Renault brand, which had a smash hit with its Logan basic car (at a price of $6,500) made in Romania in 2006, immediately started work on a future version designed to retail for around $4,000.

Goldemberg's Sweet Victory

If you still remain skeptical about the prospects for poor countries leapfrogging on energy, take a helicopter from downtown São Paulo and head north. As the seemingly unending expanse of steel and concrete below slowly yields to the green and brown and red of rivers snaking through fields, you will soon find yourself flying over an endless source of a fuel of the future.

It shouldn't be too hard to flag down a chopper, as this megalopolis now has more helicopters ferrying people back and forth over its congested streets than does Manhattan. Alas, São Paulo hasn't yet followed the example set by Singapore and

Curitiba on congestion charging, BRT buses, and other clever ways to deal with traffic. It has nevertheless managed to do something that no other big economy in the world has done: break gasoline's stranglehold. *You cannot buy a gallon of gasoline today anywhere in Brazil*: the only choices at retail pumps are pure ethanol or "gasohol" (gasoline with a minimum of 20 percent ethanol). Over 70 percent of the new cars now sold come installed with flex-fuel engines that run perfectly well on ethanol, gasoline, or any combination of the two. Visit the Zamora Volkswagen dealership in a sleepy, residential district of São Paulo, and ask the prospective buyers if they prefer the gasoline-powered car sold everywhere in the world or Brazil's flex-fuel version, and you'll find that not one customer today wants the gasoline car.

In 2006, José Goldemberg, the energy guru who has long advocated both ethanol and leapfrogging, was serving as São Paulo's state secretary for the environment. He was getting on in years, but he remained ever the feisty academic, with his tweedy jacket and a business card declaring him to be "El Profesor." "The automobile is more than a means of transportation, it's how we like to live our lives," he reflected over a *cafezinho* in his ministerial office downtown. So his leapfrog strategy has focused not on getting rid of cars per se but the mucky fuel that causes poor air quality and geopolitical complications. "We Jews joke that our people wandered forty years in the desert, and it was just our luck to end up in Israel—the only place in the Middle East without oil!"

How did Brazil solve the chicken-and-egg problem bedeviling any new fuel? Years ago, the energy guru did an important study showing that ethanol made from sugarcane (which poor, tropical countries have in abundance) can be made in a highly energy efficient and environmentally friendly way. That challenged the conventional view that all ethanol is uneconomical and ungreen. That was a reasonable criticism of American ethanol (which is made chiefly from corn, to please the midwestern farm lobby), but sugarcane is a much more energy rich crop than corn. That inherent advantage helped, but that is not the main reason Brazil has been able to break the monopoly grip of oil and become the world's leading exporter of ethanol. After all, sugarcane is grown in dozens of countries around the world, yet few have developed major ethanol industries. Even Brazil has blended some home-grown ethanol into its gasoline since the 1930s.

Brazil's success arose because of a happy collision of industry innovation, aggressive public policy, and sheer dumb luck. Brazil's agribusinesses are past masters of economies of scale. The sugarcane operations in the state of São Paulo, the country's ethanol capital, are vast, industrial-scale operations that are run like serious agribusinesses by professional managers. Most developing countries that produce sugarcane do so following a fragmented, highly inefficient, and uneconomical approach, suggesting that they too might be able to leapfrog if they learn from Brazil's modern, mechanized approach.

The government push came thanks to the 1970s oil shocks in the form of government subsidies and mandates that required a minimum ethanol content in the country's fuels. Goldemberg insists the mandate was necessary "in order to overcome the powerful resistance of the oil lobby"—a revealing comment that suggests that the Oil Curse can, in fact, be overcome by determined governments. Courage aside, this risky approach of picking technology winners was not a success at first. In fact, Brazil's initial attempt to force the market to adopt 100 percent ethanol-only cars ran into big trouble when the ethanol started to dry up. That is when serendipity came. The country made its mandates a bit more flexible, and industry in turn made car engines flex-fuel—perfecting the technology that Detroit had developed but that never really took off in America.

If poor countries like China and Brazil can challenge the petroleum orthodoxy, then surely America can too. If the BRICs turn out to be part of the solution to oil addiction and not just part of the problem, then surely the world's largest economy and its biggest gas-guzzler can come up with solutions too. Just imagine how much easier it would be to develop the clean cars and fuels of the future if the country that has come up with almost all of the disruptive business innovations and radical new technologies of the last thirty years joined the race to fuel the car of the future. Happily, that is precisely what is happening today.

CHAPTER EIGHT

The Juice and the Jalopy

*The same anarchic, amazing forces that brought us
the Internet and telecom revolutions are now racing
to develop the clean fuels and smart
cars of tomorrow*

It was midnight at the Playboy Mansion, and multimillionaires Scott Painter and Elon Musk were engrossed in an intense conversation in the cigar alcove. Their vantage point gave them panoramic views of the action and strategic access to the poolside bunny catwalks.

What could young, wealthy, and charismatic men sipping cocktails and puffing cigars at the Playboy Mansion possibly have on their minds? You would be forgiven for thinking the obvious. Painter and Musk did appear to be typical players on the scene. Painter entertained in high style at his Bel Air home, with its priceless views of the Moraga Canyon and the Pacific, while Musk loved to race around Los Angeles in his million-dollar McLaren F1 race car.

In fact, the Playboy bunnies had to struggle to get their attention that night. That is because the topic of discussion was their audacious plan to save the planet. Scott Painter has made (and lost) fortunes taking on the stodgy, centralized, and hugely inefficient car industry with new business models. Through his start-up firms Built-To-Order, CarsDirect, and Zag, he tried to bring the efficiencies and transparency made possible by modern information systems and the Internet to the manufacturing, distribution, and sales of new cars. Musk made his name selling PayPal, an online payment system, to eBay for a fortune; he is now the head of Space X, a start-up that is challenging the satellite launch oligopoly by offering private space launches.

That night, Painter was arguing that his early failed attempts at overthrowing Detroit were simply the result of being too far ahead of his time. Several industry trends have converged, he argued, to make it much easier for start-up firms to challenge the big automakers today. First, he pointed out, the key intellectual property involved in making cars is no longer guarded in house by the likes of Ford and GM: they outsource most aspects of new cars (except engines) to global parts suppliers, integrators, module manufacturers, and so on. That makes it much easier for newcomers just to buy that know-how off the shelf from the suppliers, including vital knowledge on making cars safe. Second, he observed, the cost to launch a new car company has dropped dramatically. When Toyota and GM launched, respectively, Lexus and Saturn as semi-independent new companies, they had to spend billions of dollars. Now, Painter argued, an upstart now would need just a few hundred million dollars: "Never before have the barriers to entry been so low in cars!"

Elon Musk agreed, though for different reasons. He pointed out that it was still not easy to take on the incumbents: "The last successful car start-up in America was a hundred years ago." Even so, he was convinced that the time had come because of a fundamental "technology discontinuity": the shift from the century-old internal-combustion engine to electric drive. The proportion that electronics makes up of a car's cost has shot up from tiny to close to a quarter, and by 2010, experts think it could approach half. Thanks to dramatic recent advances in batteries and power electronics (made possible by the cell phone and laptop industries), Musk thought cars will go all the way to full electric: "In fifty years, we'll look back on the internal-combustion engine and see it as a giant anachronism, like the steam locomotive." Such a paradigm shift, he argued that night, would at long last break the conventional motor industry's grip and shift power to rivals that have competitive advantages over the Rust Belt in making sophisticated electronics: the nimble, New Economy companies of Silicon Valley (or for that matter, Singapore or even China).

A few days later, Elon Musk proved that he is a man who puts his money where his mouth is. At a launch party held inside a hangar at Santa Monica Municipal Airport, he unveiled the all-electric Tesla Roadster. When you say electric car, most people have an image of golf carts and other entirely unsexy vehicles. Tesla Motors aims to alter that perception. The venture, based in California and run by Musk (with money from Larry Page and Sergey Brin, the cofounders of Google), unveiled a smoking-hot, two-seat sports car. It costs just under $100,000, and Tesla aims to sell a couple thousand of them before introducing a cheaper four-seat version.

The car's design alone is likely to turn old-fashioned notions of electric vehicles on their head. Beyond that, Tesla makes three daring claims. The first is that the vehicle accelerates from 0 to 60 miles per hour in just four seconds. That is faster than a Ferrari. The second is that it can travel 250 miles on an overnight charge from a household 240-volt socket (not the special public charging stations that the earlier generation of electric cars required). The third is that it is more environmentally friendly than a gasoline-powered equivalent.

There is no doubting its breathtaking quickness, as the white knuckles and palpitating hearts of those lucky enough to test out the new roadster that evening in Santa Monica can attest. And the range of 250 miles is a heroic accomplishment, double or more that achieved by the earlier generation of electric cars. This is made possible by the use of advanced lithium-ion batteries and lightweight carbon-fiber bodywork. Using the official government methodology, the car's fuel efficiency is the equivalent of 135 miles per gallon of gasoline.

The environmental claim might sound a bit fishy to some. Yes, electric cars emit zero local pollution, but surely one must take account of the fact that dirty fossil fuels are used to make much of the electricity used by such cars when they recharge at night—a fact that has prompted wags to call them "pollute somewhere else" cars. Even taking that factor into account, it turns out the green claims are justified. Serious studies have shown that modern electric vehicles that draw their power from a grid that is itself half coal fired (as America's is) produce less in the way of greenhouse gases than an average gasoline-powered car. And as the grid "decarbonizes" over time, thanks to carbon regulations and the rise of low-carbon technologies like renewables and carbon sequestration, the advantage electric cars have over gasoline cars will only increase.

Tesla, though, aims to be even greener than that. The firm plans to offer optional solar-photoelectric systems for a few thousand dollars, to be set up as a carport at home, that will be able to power the cars for perhaps 50 miles a day without having to draw on the grid. Given that the average American drives only 25 or 30 miles a day, the idea promises, as Musk puts it, to "make our cars energy positive." You could actually sell excess green power back to the grid even as you put Ferraris to shame on the freeway!

The crowd of car connoisseurs that gathered at the Santa Monica event still had one gripe, though. Some of the gearheads and racing purists complained that the silence of the electric motor was alien. They missed the grunt and growl of an internal-combustion engine. A Tesla engineer nearby came back with an idea: "We'll program the software to have a variety of engine roars, just like ring tones on mobile phones."

The arrival of the Tesla captures vividly the twin pathways of alternative fuels and alternative vehicle technologies that are upending the world of oil and cars. Simply put, both the juice and the jalopy are undergoing radical transformation. On the one hand, alternative fuels are challenging gasoline's grip. Efficiency, electricity, ethanol, and hydrogen can all be seen as rival "fuels." However, alternative fuels will not be enough to break the world's addiction to oil, given petroleum's power of incumbency. Indeed, there is reason to think that a well-intentioned but partial move to clean fuels will only help Big Oil, which is cleverly finding ways to blend those supposedly alternative fuels right into conventional gasoline and diesel.

That is why radical change requires jazzing up the jalopy too. The very basics of the automobile are changing, from a grease-and-grime, stamped-steel approach to one that relies much more on silicon chips, lightweight carbon fiber, smart technologies, and software, turning the car into the ultimate electronic accessory. As this happens, Scott Painter's argument about lower barriers to entry becomes suddenly relevant, since the world's centers of excellence for electronics and advanced composite materials lie not in Detroit but in Singapore, at Sony, and, of course, all over Silicon Valley.

The arrival of the Tesla proves that the industry is now open to breakthrough innovations. However, that is not to say that electric cars are sure to replace gasoline ones. The field is wide open today, with a range of alternative fuels and engines under development, and the winner is far from certain. Indeed, there may well be a variety of successors to today's dominant gasoline-burning internal-combustion engine. The global race to build and fuel the car of the future is on, just as it was a century ago, when gasoline, biofuels, and electrics first raced for supremacy.

Gentlemen, Start Your Engines

In 1894, *Le Petit Journal* of Paris organized the world's first endurance race for "vehicles without horses." The race was held on the 80-mile route from Paris to Rouen, and the purse was a juicy 5,000 francs. The rivals used all manner of fuels, ranging from steam to electric batteries to compressed air. The winning car used a Daimler engine fired by a strange new fuel that had previously been used chiefly in illumination as a substitute for whale blubber: oil.

Despite the victory, petroleum's future seemed uncertain back then. Internal-combustion vehicles were seen as noisy, smelly, and dangerous. Indeed, the turn of the century was probably the golden age of the electric car. In 1889, Thomas Edison built the first real electric car, with a rechargeable battery. In 1896, the nation's first car dealer was selling only electric cars. Two years later, when a spectacular blizzard buried New York City under more than 3 feet of snow, only electric cars could move on the roads. New York's prime source of transportation, horse-drawn vehicles, was dumping 2.5 million pounds of manure and 60,000 gallons of urine daily on the streets. About fifteen thousand dead horses had to be dragged off the streets each year.

By this time, a third of all cars were powered by steam, one-third by electric, and one-third by gasoline. In 1908, even Henry Ford bought his wife an electric car.

Why did gasoline emerge victorious? Some conspiracy theorists believe it was the result of devious tactics employed by the oil and battery companies of the age. Edwin Black, an investigative journalist, makes that case in his book *Internal Combustion*. He notes that even Henry Ford began to think that electric cars were better than gasoline ones, joining hands with Thomas Edison to mass-produce electric cars. Alas, that joint venture never came to fruition. Black points an accusing finger at various possible conspiracies, oligopolies, and dirty dealings that might have killed that effort. Black is probably right about vested interests fighting dirty to protect their turf back then—much as Big Oil and Detroit do today—but one does not need elaborate conspiracy theories to see why oil triumphed.

One factor was Spindletop, the raging oil field in Texas. The discovery of plentiful oil in Texas at the turn of the century helped push petroleum and gasoline to the fore. The invention of the electric starter-motor, in 1912, also tipped things in gasoline's favor. By eliminating the arduous and dangerous starting handle, the electric starter overcame the biggest obstacle to the widespread adoption of the gasoline engine, as it suddenly became easier to use. Soon gasoline's inherent superiority as a motor fuel caused it to prevail. Gasoline has such high energy density that it offered power and durability that electric batteries and coal-fired steam engines could not match using the technology of the day.

Pretty soon Henry Ford was churning out Tin Lizzies by the thousands. His cars were lightweight and basic stripped-down vehicles that ran on gasoline or ethanol. Since then, cars have grown heavier and bigger. But what goes around comes around.

Not only are flex-fuel cars that run on gasoline and ethanol back in fashion, but electricity is making a comeback.

Reduced Dependency or Risky Dead End?

Believe it or not, diesel is coming out of the closet. Even in America, where *diesel* is a dirty word, heads turned when an Audi sports car running on Shell diesel won the Le Mans endurance race in 2006. Before that, the youthful Formula 1 ace Jenson Button, too young to rent a gasoline-powered car in Europe, was caught speeding by the French police doing 140 miles per hour in a BMW diesel. Most Americans did not know that diesels could go so fast. For Americans, diesel is associated with heavy trucks, smoky buses, and massive railroad locomotives. But the continental European experience has been different and is now coming to the United States.

European governments first started taxing diesel more lightly than gasoline in the mid-1970s, after the first oil shock. The aim was to limit oil imports, which were straining their weakened economies, already suffering high inflation from the knock-on costs of the quadrupling of the oil price. Since then, technical developments such as efficient turbochargers and sophisticated direct electronic fuel-injection systems have vastly improved the performance of diesels in ordinary sedans and SUVs. Since 2000, Mercedes joined BMW in emphasizing diesel for performance as well as economy, rather than offering it as a reliable, economical standby for German taxicabs.

In the United States, 97 percent of cars run on gasoline, but across the Atlantic, the picture is very different, thanks to an important lesson learned by European governments after the oil shocks of the 1970s. Today, half of all European cars sold are diesel-powered, and in the upscale segment populated by Mercedes and BMW sedans, the proportion is 70 percent, pushed up there by vehicle taxes that favor diesel. While engine technology has advanced, so has the exhaust treatment to reduce harmful emissions. Because diesel combustion takes place at a lower temperature than gasoline combustion, the exhaust contains more of the noxious nitrogen-oxide gas that combines with sunlight to form smog. The pollution problem caused by emissions of nitrous oxides has kept diesels from succeeding in America, where many still remember the smoky and sooty exhaust belching forth from the tailpipes of Audi and Mercedes cars from the 1970s.

But now Mercedes thinks it has that problem licked with a new technology known as Bluetec. The Bluetec system uses a combination of technologies to dispel the pollution of old-style diesel autos. Mercedes claimed when it launched the technology in Detroit in January 2006 that such diesel engines offer fuel economy that is between 20 percent and 40 percent better than gasoline engines. J. D. Power, the consumer testing and research firm, has forecast that by 2015, more than one in ten American cars will be diesel.

But even if diesel does not do it for American drivers, gasoline technology is not standing still when it comes to improving technology. Basically, the dividend comes in two forms: better gas mileage or better performance from the same consumption, more bang or less buck, whichever way drivers want it. BMW has a more refined form of direct-injection gasoline engine that involves a continuous flow of fuel and air mixture, rather than the conventional stop-start delivery, resulting in claimed improvement in economy of up to 15 percent. But the biggest potential breakthrough will be the arrival on the road of a type of gasoline engine that has been discussed for years in the technical press. It is called an HCCI engine: homogenous charge compression ignition.

Honda, Ford, GM, and some other manufacturers, including Mitsubishi, are working on this. In effect, HCCI engines are gasoline-fueled diesel engines in that they use pressure rather than spark plugs to ignite the mix of air and fuel. By copying the diesel engine's operating cycle, they achieve a similarly superior thermal efficiency and, hence, fuel economy. The outcome is about 80 percent of a diesel's efficiency for half the cost, since the engine block does not have to be as massive as it does for a diesel. GM reckons HCCI also delivers lower emissions. DaimlerChrysler is working on another variant of diesel technology with an engine that runs on a hybrid mixture of diesel and gasoline. The aim is to combine the cleanliness of gasoline with the miles per gallon of diesel.

There are other advances coming down the pike. Fiat, the European car company that always seems to punch above its weight in terms of technology (it invented the common rail direct-injection technology that transformed diesels), has an Alfa Romeo with a new kind of fuel injection being tested. Instead of using conventional camshafts to control the valves that let fuel into and exhaust gases out of cylinders, it has inlet valves actuated electrohydraulically. This allows variable valve timing that is much more precise and avoids the pumping losses associated with a normal aspirated throttle, when the driver floors the pedal. The reduction in mechanical friction, the separate value timing, and the lighter engine (minus camshaft) can boost efficiency by some 20 percent. Mercedes, Lotus (a small specialist company in Britain owned by Malaysian interests), and other companies are working on similar electropneumatic valve technology.

All these developments in diesel- and gasoline-burning engines point to several powerful but usually overlooked insights. First is the fact that efficiency gains, such as those made possible by diesel engines, are helpful to a point—but risk being a dead end. Using petroleum more efficiently is obviously a good thing, but if better diesels and smarter gasoline engines are developed instead of rather than in tandem with more disruptive inventions and breakthrough technologies, then the world will remain addicted to oil ten or twenty years hence, with no way out. In other words, better juice is not enough if the jalopy doesn't change too.

The second insight, and another reason for clean-energy innovators to redouble their efforts, is that the incumbent never stands still. As transportation system analysts at MIT have pointed out in various studies looking at the future of automotive technologies, tomorrow's internal-combustion engine will be cleaner, smarter, and more efficient than today's. Indeed, their own engineering colleagues are coming up with some of those innovations, as this eye-opening press release from late 2006 made clear:

> *MIT researchers are developing a half-sized gasoline engine that performs like its full-sized cousin but offers fuel efficiency approaching that of today's hybrid engine system— at far lower cost. The key? Carefully controlled injection of ethanol, an increasingly common biofuel, directly into the engine's cylinders when there's a hill to be climbed or a car to be passed.*
>
> *These small engines could be on the market within five years, and consumers should find them appealing: By spending about an extra $1,000 and adding a couple of gallons of ethanol every few months, they will have an engine that can go as much as 30 percent farther on a gallon of fuel than an ordinary engine. . . . If all today's cars had the new engine, current U.S. gasoline consumption of 140*

billion gallons per year would drop by more than 30 billion gallons.

There is a lot of hoopla surrounding alternative fuels today, but caution is in order. For one thing, no single alternative fuel can displace the entire petroleum economy: we will need all of them, probably, working together in a clean-energy portfolio of the future. Even then, it will not be easy to change the current paradigm unless there is a fundamental transformation of the automobile itself—as MIT's cheeky use of ethanol to boost the power of a gasoline engine shows.

The Biofuels Bandwagon

It must rank as one of history's more unlikely conversions. President Bush is an oil man from Texas, but in his 2006 state of the union speech, he declared that America is "addicted to oil" and trumpeted the virtues of ethanol, a motor fuel made from grain alcohol. Bush says he wants a vast expansion of the country's tiny ethanol industry. In particular, he wants cellulosic ethanol prepared using an advanced technology, which he argued would become commercial within six years. Will it happen? Ethanol will not replace oil anytime soon, but Bush nevertheless has put his finger on something big. This once-sickly, oversubsidized industry is brimming with so much optimism and investment that there were even signs of a speculative financial bubble in ethanol in 2007.

America has traditionally made ethanol from corn. This is much less efficient than Brazil's sugarcane ethanol, but corn has the politically important advantage of being grown in the midwestern states. Ethanol is blended with 15 percent gasoline to form so-called E85. In some states, thanks to hidden subsidies in the form of tax waivers, it has sometimes been cheaper than gasoline, as when oil prices of more than $60 a barrel pushed pump prices up to around $3 a gallon.

There are already five million flex-fuel cars in the United States with engines tuned to run on either gasoline or ethanol, just like Henry Ford's Model T. For years, these have been a running joke in Detroit, as carmakers made them only to collect bogus credits that would let them off easy on fuel-economy standards. Many of those millions of customers did not even know their cars were flex-fuel. But the political push by President Bush and the broader biofuels boom has changed that. GM trumpeted Detroit's newfound seriousness about flex-fuel cars with a splashy advertising campaign launched during the 2006 Super Bowl encouraging drivers to go green by "going yellow" (as in the color of corn).

In 2006, GM, Ford, and Chrysler pledged to double their output of such vehicles, from a million a year, by 2010. If all these vehicles were to run on E85 (85 percent ethanol and 15 percent gasoline), it has been estimated that the United States could save more than 3.5 billion gallons of gasoline a year. Of course, it is one thing to have millions of flex-fuel cars running around and another to actually fill them with ethanol or biodiesel instead of regular fuel. Most simply consume gasoline for the simple reason that ethanol is hard to find on the road. Barely 700 of America's 170,000 gas stations offer ethanol. Even so, sales of ethanol have been growing at a staggering 30 percent a year recently.

Advocates have high hopes. High oil prices, government support, and the promise of new technology have led to a veritable boom in production of American-style ethanol. Several billion dollars of investment, led by agribusiness giants such as Cargill and Archer Daniels Midland, are going into new production plants for corn ethanol. Daniel Kammen of the University of

California at Berkeley thinks the future does not belong to corn ethanol, which takes a lot of energy and subsidies to make, but nevertheless argues that it's not as bad as it seems. Using it releases less greenhouse gas than does burning gasoline, he calculates, and its spread helps develop infrastructure for the much greener ethanol that should come onto the market in a few years' time.

Other people have their hopes pinned on biodiesel, a different sort of biofuel that can be blended into conventional diesel fuel. Willie Nelson, the country singer, drives around in a Mercedes powered by his own brand of biodiesel. BioWillie is made from vegetable oil, but rival biodiesel blends use soybeans, rapeseed, switchgrass, and even garbage and turkey carcasses.

Biofuels add up to barely a few percentage points of America's transportation fuels today, but states are hoping that will change. In Montana and Minnesota, gasoline must contain 10 percent ethanol, while Iowa's laws require that by 2020, fully a quarter of all fuel sold in the state be gotten from renewable sources. And Congress has mandated a near doubling of biofuel production by 2012. One advantage biofuel has over hydrogen, another contender for the fuel of the future, is that it does not immediately need new infrastructure to support its rollout. Ethanol can be blended into gasoline in lesser concentrations and used in today's normal engines. But for biofuels to make a real dent in oil's monopoly, we will need a nationwide network of dedicated filling stations and infrastructure for supplying consumers fuel blends with high concentrations of ethanol. Getting to such a vast network of filling stations will not be easy.

Leapfrogging Out of Big Oil's Tentacles

This points to an important risk with biofuels that hydrogen does not suffer from (though it has plenty of obstacles of its own to overcome): precisely because they can be blended into gasoline and folded into the oil infrastructure, there is a possibility that the current boom in biofuels will end up perpetuating the world's addiction to oil rather than ending it. That fear is bolstered by evidence on the ground that Big Oil is even now gearing up to embrace biofuels in a modest way so that it can "manufacture" the gasoline of the future.

Big Oil (in the shape of BP) and Big Chemistry (in the shape of DuPont) and Big Agribusiness (in the shape of British Sugar) have developed another biofuel known as biobutanol, which has the advantage of fitting in well with both the current oil-distribution system and conventional gasoline engines. Cynics worry that this is like feeding a heroin addict a mixture of the drug and methadone rather than making the switch to the more manageable surrogate. In other words, Big Oil may be moving in to keep us all hooked on oil for our automobiles while posing as a Big Green Giant. This is just one of the mechanisms that cut in when oil prices are high and when alternative fuels creep into the limelight, offering promises of relief.

Even greens, who generally like biofuels, have some concerns. Nathanael Greene, of the Natural Resources Defense Council, is a big booster, but he is careful to emphasize that biofuels are not "a silver bullet." In a comprehensive report, his group argues that American output of biofuels could soar from puny levels to 100 billion gallons by 2050. But even then, it would not dislodge oil. Other greens worry about the pesticides involved in a big scale-up of ethanol production, as well as the soil erosion likely to result. Those watching the rain forests of Malaysia and Indonesia being chopped down to make way for palm-oil plantations also worry about biofuels: that monoculture crop is an increasingly popular way to make biodiesel.

The Great Green Hope

Most, though not all, of these objections could of course be overcome if the technology breakthroughs that boosters now see in the laboratories and at pilot biofuels plants actually scale up in the commercial marketplace. In particular, the biofuel to watch is not corn ethanol or biodiesel made from garbage but cellulosic ethanol. If perfected, this newfangled ethanol would be not only more efficient and greener than even Brazilian sugarcane ethanol (which is already pretty green and efficient) but it could also be made from virtually any agricultural material—be that prairie grass or agricultural waste—and not just corn. Cellulosic ethanol also offers a convenient solution to the political impasse over ethanol. The farm lobby's power means that America doles out billions of dollars in subsidies to producers of corn ethanol, which clearly has no future and can never hope to challenge oil—and unfairly and perversely imposes tariffs on imports of the greener, cheaper Brazilian variety.

The best reason for optimism about next-generation biofuels is the arrival of entrepreneurial capital. Big multinationals have been pouring money into this sector for a while, of course. Royal Dutch/Shell has a joint venture with Canada's Iogen, which plans to open a commercial plant by 2009. DuPont and Genencor, a biotechnology firm, are also busy developing better catalysts to make the production process more efficient. But the Silicon Valley money is new and its controllers infectiously optimistic. Paul Allen and Bill Gates, cofounders of Microsoft, have both made recent (but unrelated) investments in ethanol firms. Richard Branson, the British airline boss, has jumped into the fray with Virgin Fuels, a new firm that vows to invest $3 billion in ethanol and related clean fuels over the next few years. Vinod Khosla, a venture capitalist at Kleiner Perkins Caufield

& Byers, has also put his own money into start-up firms developing cellulosic ethanol.

Strikingly, Khosla speaks with the zeal of a convert: "I have a religious belief in the power of ideas propelled by entrepreneurial energy. Bush is too cautious, cellulosic ethanol can take off much sooner than six years!" he declares. Coming from some businessmen, such talk might sound self-serving or nutty. But Khosla helped to found Sun Microsystems, a company that pioneered such essential bits of Internet technology as network servers and Java, a programming language. He then made his name and his fortune as a partner at Kleiner Perkins, a Silicon Valley venture-capital firm famous for its early investments in AOL, Amazon, Compaq, and Google. His eyes have now turned toward a new target—the oil industry. Anyone who spends time with him is liable to be hit with his well-researched but mind-numbing PowerPoint presentation on ethanol—unveiled with the affection that some men reserve for pictures of their grandchildren.

Why is Khosla embarking on this particular crusade when he could concentrate on the technology investments that have served him so well—or even opt for a gilded retirement? Like many very rich men, he now wants to improve the world: "Just starting another Sun doesn't do it for me anymore." As an engineer turned venture capitalist, Khosla has a healthy respect for the power of new technologies to create disruptive innovations. And the free marketeer in him clearly relishes the prospect of really taking on the big, rich, and well-entrenched firms that dominate the oil industry.

Another part of the explanation lies in his complex relationship with India. Like several of Silicon Valley's most successful people, Khosla boasts a degree from the Indian Institute of Technology. When he tried to start a project to help the mother country, he was initially frustrated by its bureaucracy and corruption. His first attempt to start a traditional top-down charity failed, so he now funds only charities embracing microenterprise approaches. A lesson he learned from India, he says, is that one has to think big: "Unless you influence the lives of at least a million people, it simply doesn't matter." His plan is to use technology and entrepreneurship to tackle big social and environmental problems: "In venture capital, we fail far more often than we succeed," he says. "I've decided that I'd better focus on taking on problems that really matter, so that when I win it makes a difference to the world." He likens his need to get involved with worthy causes to a drug addiction.

It is easy to dismiss this enthusiasm as the irrelevant obsession of a rich hobbyist or the harmless utopianism of a capitalist who has made his pile. But the big oil companies are certainly not taking Khosla lightly. They are spending lavishly to counter his lobbying on behalf of ethanol. Perhaps the best reason to take Khosla seriously is that his professional success and Republican leanings mean that he has the ears of powerful people. He has been making the rounds, from the White House and Capitol Hill to the World Economic Forum at Davos and the Technology, Entertainment and Design (TED) Conference (a big annual gathering for the leading lights of those three industries), banging the drum for ethanol. Before Larry Page, Google's cofounder, attended a recent TED conference in Monterey, California, he was skeptical about ethanol. After hearing Khosla, he decided to help fund the cause. "When have you ever seen greens, farmers, and guys like me and Larry on the same page?" demands Khosla.

The Power of Price

Seeing people like Khosla investing in biofuels has certainly sparked a boom. Such jubilation is understandable, but it may be slightly premature for two reasons: the possibility of oil prices dropping sharply, as they did in 1986, 1999, and again in 2005; and the inconvenient fact that high oil prices help dirty "alternative" fuels as much as clean ones.

Green investors are unwise to rely on permanently high oil prices. That vast reserves of oil remain in the hands of a few autocratic and possibly unreliable regimes in the Persian Gulf suggests that volatile, not reliably high, prices will be the norm. And as the earlier, failed renewables booms of the 1970s and 1980s showed, oil-price decreases can wipe out alternative-energy schemes. Technological breakthroughs and green policies like instituting carbon taxes suggest that this renewables boom may be more sustainable than the last one. But investors counting on sustained high oil prices to justify otherwise uneconomic projects should beware.

Even so, Khosla is convinced his grand plan will work. He insists "this fuel is greener, cheaper, more secure than gasoline— and this shift won't cost the consumer, automakers, or the government anything." There are undoubtedly attractive features of ethanol. But making the switch may not be as easy or cheap as Khosla suggests. Retail distribution is one obvious problem: expanding infrastructure will cost money and take time, and the oil industry is not exactly enthusiastic about it. And cellulosic technology, which seems so promising today, may take much longer than expected to achieve commercial scale or might fail altogether.

What is more, the OPEC cartel is suspected by some of engineering occasional price collapses to bankrupt investment in alternative energy. Khosla himself says that after he made his ethanol pitch at a recent Davos summit of world leaders and corporate titans, a senior Saudi oil official sweetly reminded him that it costs barely a dollar to lift a barrel of Saudi oil out of the ground, adding, "If biofuels start to take off we will drop the price of oil."

Anticipating this problem, Khosla is lobbying politicians in Washington, D.C., to impose a tax on crude oil if the price falls below $40 a barrel to safeguard investments in ethanol. He has also proposed reforming the ethanol subsidy regime so that it drops when oil prices are high and kicks in when oil prices are low—thus acting as a buffer against an oil-price collapse.

There is a second reason to regard the recent jubilation among green investors with skepticism. Clean energy alternatives like biofuels are not the only sort of "alternative" energy that is enjoying a boom: dirty technologies like Canada's mucky tar sands (from which gasoline can be made at great environmental expense) are also benefiting from high oil prices. In theory, there is as much energy trapped in Alberta as in all of Saudi Arabia. In practice, it has proved too complex and expensive to be a serious rival to oil—until now.

Today's oil prices, combined with cost reductions and innovations in tar-sands processing, are leading to a bonanza. Peter Tertzakian of ARC Financial, a Canadian investment firm, estimates that the flurry of investment in new projects and expansion efforts in the processing of tar sands will add up to a whopping $60 billion or so in coming years.

To the chagrin of greens, today's high oil prices are giving even filthy coal, their bête noire, a new lease on life. Several American utilities are now talking of building new coal-fired power plants, partly to hedge against the natural-gas price risk. More striking is a coal deal recently announced by the State of Pennsylvania and a consortium that includes the oil giant Royal

Dutch/Shell. Using an innovative process (developed by South Africa's Sasol), the consortium will convert coal waste into a liquid that can be blended into normal diesel fuel.

The rise of coal, tar sands, and shale shows that a market surge in oil prices (as opposed to higher oil prices resulting from carbon taxes) is not an unambiguous win for the greens. It also proves that the promising move to alternative fuels like ethanol is still a tenuous one, given the ability of the incumbent powers of oil and cars to co-opt them and perpetuate the life of their existing investments. This is not a preordained outcome: Vinod Khosla rightly points out that merely having existing infrastructure is no guarantee that nimble newcomers with lower marginal costs won't eat your lunch. He helped bring about precisely such revolutions in various bits of the software, Internet server, and computer industries and saw firsthand how shortsighted the dinosaurs could be. That is why it is worth taking a closer look at the seismic shift now taking place in the oil industry thanks to Big Oil's new strategy of manufacturing gasoline.

Manufacturing Gasoline

The global oil industry is on the verge of a dramatic transformation from a risky exploration business to a technology-intensive manufacturing business. And the product this industry will soon be manufacturing, argues Jeroen van der Veer, Shell's chief executive, is "greener fossil fuels."

To see what he means, consider the surprising outcome of another great motorcar race. In March 2006, at the Sebring test track in Florida, a sleek Audi prototype R-10 became the first diesel-powered car to win a LeMans-style endurance race, beating a field of gasoline-powered rivals to the finish line. What makes this tale extraordinary is that the diesel used by the Audi was not made in the ordinary way, exclusively from petroleum. Shell blended a superclean and superpowerful new form of diesel manufactured from natural gas (with the clunky name gas-to-liquids, or GTL) into the Audi's tank. Several big GTL projects are now under way in Qatar, whose North Gas Field is the largest in the world. Nigeria and others seem likely to follow.

Since the world has far more natural gas left than oil, manufacturing fuel in this way would greatly stretch the world's remaining supplies of oil. So too would blending gasoline or diesel with ethanol and biodiesel made from agricultural crops, as Brazil does today, or with fuel made from Canada's tar sands or America's shale oil. Using technology invented by Germany's Nazis and perfected by South Africa's Sasol during periods when those countries were under oil embargo, companies are now also investing furiously to convert coal into a clean liquid fuel that can be used in conventional car engines. All of these advances, argues the oil expert Daniel Yergin, mean that "the very definition of oil is changing, since nonconventional oil becomes conventional over time."

One reason that Big Oil companies are investing in these alternatives is undoubtedly price. There is a growing sense among oilmen that oil prices will not return to the $20s, the level that prevailed through much of the 1990s, never mind the $10 price reached in 1999. Because many of the current alternative fuels are economical when the oil price is above $40 a barrel, argues Peter Robertson, vice chairman of Chevron, "price is our friend" in debunking the misguided argument that the world is about to run out of oil.

Ah, but what if there is a price collapse? If supply remains robust or global demand cools off, there might be a sharp (and largely unexpected) drop in oil prices. Indeed, OPEC could engineer such a price collapse to wipe out alternative energy, as it has done in the past. Veteran energy observers believe that American officials put pressure on the Saudis back in the Reagan era to engineer the big price collapse of 1985 that they believed (perhaps rightly) would lead to the economic collapse of the oil-rich Soviet Union. Keen to prevent another OPEC-induced price collapse, Richard Lugar, a Republican senator keen on ethanol, is now promoting a radical idea: an oil tax. His tax would kick in only when oil prices fell below $35 a barrel and so act as a floor price for the oil market.

If the oil price collapsed again, it would undoubtedly wipe out a good deal of investment in alternatives—but not all of it. There is much money today going into absurd and uneconomic ideas like making oil out of turkey gizzards and biodiesel out of used french-fry fat. That sort of "alternative" fuel would surely fall by the wayside—and deserve to.

However, technology advances are rapidly bringing down the cost of GTL, tar sands, cellulosic ethanol, and several other more viable alternatives. Big Oil is likely to push ahead with investments in these areas even if prices drop for a while. The reason is not because oil companies are "beyond petroleum" already, their advertisements not withstanding. No, it is because Big Oil fears lack of access to conventional oil reserves in the Middle East and Russia. Because the big Western firms are having difficulty replacing the oil reserves on their books, they are grasping at any and all fuels that can be blended right into their existing fuel.

If you doubt the oil industry's determination to squeeze every last drop out of the oil economy, visit the Kern River oil field near Bakersfield, California. This supergiant field is part of a cluster of oil fields that has been pumping out black gold for over one hundred years—Howard Hughes made his family fortune on one of these fields. Kern River has produced over two billion barrels of oil and has perhaps another two billion left. The problem is that it is extremely heavy oil that is very difficult and costly to extract. Chevron has invested billions of dollars in a sophisticated steam-injection process that has not only kept the field going but has dramatically increased its output in recent years. Through massive automation (each engineer looks after a thousand small wells drilled into the reservoir), the firm has transformed a process of "flying blind" into one where the well "practically monitors itself and calls for help when it needs us."

This is not a freakish anomaly. China has vast deposits of heavy oil that would benefit from such an advanced approach. The United States, Canada, and Venezuela have deposits of heavy hydrocarbons that would dwarf even the Saudi oil fields. And the Saudis have recently invited Chevron to apply its sophisticated steam-injection techniques to recover heavy, previously uneconomic oil in the neutral zone the country shares with Kuwait.

Unconventional fossil fuels are, simply put, a potential carbon nightmare. Alex Farrell of the University of California at Berkeley calculates that if just a quarter of the world's coal is converted to motor fuel via the coal-to-liquids process, the greenhouse-gas impact would be bigger than burning all of the world's conventional petroleum. And that does not even begin to talk of the climate impact of manufacturing gasoline from tar sands, shale, and other unconventional hydrocarbons.

Those vast, dirty reserves explain why the oil will not run out for a very long time and why alternative fuels by themselves do not have a chance of dislodging petroleum from its throne. To get to clean, carbon-free transport, the juice needs the jalopy to change too.

Electrifying the Jalopy

So if there are no wonder solutions from alternative fuels, what about fantastic new engine technology that produces as much power but offers much better gas mileage? Better diesels and internal-combustion engines are going to be contenders for a while, but these are essentially incremental technologies that defend the vested interests of oil. The potential game-changers are flexible engines, hybrid electrics, and fuel cells, and the common thread to all three is that they are made possible only through big advances in power electronics, control software, and sophisticated electronic hardware. In other words, the jalopy of the future is going to be wired (or, if you have a satellite hookup, your ultimate wireless accessory).

Thirty years ago, electronics accounted for only about $110 of the cost of the materials in a car, around 5 percent. Today, at $1,400 to $1,800, various electronic devices comprise up to 20 percent of the materials cost. Industry experts reckon that 80 to 90 percent of innovations in automobiles nowadays relate to electronics, something that Mercedes learned bitterly a few years ago when its new E-Class models were dogged by breakdowns and recalls because of electronics faults. Electronic engine controls have been behind the big gains in gas mileage that have come along in the past twenty years. The traditional carburetor so beloved of gearheads and home mechanics was little more than a bucket for pouring gas into each cylinder. Today's finely calibrated injection systems are like space rockets compared with the Wright brothers' early flying machines.

But electricity has been creeping up. Volkswagen and Audi fiddled around with hybrid gasoline-electric cars in the 1980s. There was a version of the Audi 100 Avant that had a gasoline engine driving the front wheels and an electric motor at the back. It required a huge, heavy battery pack and was sluggish when run on electric power. But a huge step forward was made when Toyota brought the first Prius to market in 1997. Since then, its fuel economy has risen from 40 to 55 miles to the gallon while its acceleration has improved by a third, so that the second-generation Prius can now go from 0 to 60 miles per hour in ten seconds. Meanwhile, its emissions have fallen by 30 percent. Toyota halved the cost of the hybrid production system in the first six years and is confident of doing the same again with the latest model. Despite costing up to $4,000 more than conventional cars of a similar class, the Prius proved a hit, especially in California and Japan.

The first of the game-changers is flex-fuel technology. Brazil has already shown how powerful flex fuels can be in breaking the old chicken-and-egg problem of fuel and engine. The first time Brazil pushed ethanol, the lack of flex-fuel engines forced buyers to commit to 100 percent ethanol engines; when the clean fuel got scarce, they were stranded, and their cars became worthless. Now, flexible engines are so wildly popular that car-makers cannot seem to sell gasoline-only cars.

Ethanol does not have a monopoly on flexibility, though. BMW has developed an internal-combustion engine that runs on either gasoline or hydrogen fuel. Such an engine does not use hydrogen in a perfectly zero-emissions way, since it burns the hydrogen; even so, the resultant emissions are so low that they meet the toughest "super ultra low" emissions standards set by California regulators. The ultimate internal-combustion engine of the future will probably burn either diesel or gasoline as a backup fuel, while its plug-in hybrid feature will allow you to toggle effortlessly and efficiently between electricity and another clean fuel of your choice. The future may well belong to flex-fuel, perhaps even poly-fuel, engines simply because it is not yet clear which fuels will emerge victorious in the future, and of course nobody wants to be stranded.

Hybrids are the second game-changer. The most important thing to remember about hybrids is this: despite all the hoopla, the Toyota Prius is not the car of the future. However, like flexible engines, the hybridization of cars is an obviously good idea and a clever way to hedge bets in a risky world of changing fuels. Indeed, one could argue that, as the pioneer of hybrid technology, the Prius is the essential stepping-stone to all future cars, regardless of whether those are ethanol-reliant, plug-ins, or hydrogen fuel cells.

What the hybrid does is offer a huge boost to the practical efficiency of the internal-combustion engine. It extends the scope of what it offers. It's like the tonic in a gin and tonic. So it plays to the energy-conservation argument, making the petroleum last longer, go further. The downside is that such developments prolong oil addiction, though they make it easier to live with it. The upside, indeed the real significance, of the hybrid is that it enables all the incremental improvements to internal-combustion engines to be rolled up into it and harnessed, while its boost to the development of lighter batteries and more compact, lightweight control electronics systems prepares the way for fuel-cell–electric cars in the future.

The hydrogen-fuel-cell car will be a hybrid in itself, the car of the future, employing the technology that nearly a million Americans are driving today. So the gasoline-electric vehicle that has been such a surprising success in America and Japan is absorbing today's technology advances while forming a bridge to the required technologies of the future. It is simultaneously offering an extension of the Age of Oil while helping to pave the way for the postpetroleum era of the hydrogen economy.

So How Does the Mojo Work?

Toyota and Honda are way ahead of other car companies in developing hybrids and in their mass production. The likes of Ford, Nissan, VW, and Peugeot have to license Toyota technology or else, like GM, DaimlerChrysler, and BMW, form a consortium to attempt to catch up. By the end of 2006, the Japanese pair had brought production costs down low enough to make profits on the vehicles, even though they cost about $4,000 more to manufacture and sell over the price of a comparable sedan. Every other car manufacturer has been obliged to introduce hybrid versions of normal models in an attempt to siphon off some sales and some green glory from the Japanese pioneers.

Toyota responded to the surprise acceptance of the Prius in America by starting production in its Kentucky plant. Its first Prius plant outside Japan was in China, where it wanted to take advantage of low labor costs for making cars for export while impressing the Chinese government with its willingness to bring the latest technology into China, a country increasingly looking to the technological future when it comes to mass motorization.

While the mass-produced hybrid is new, the idea goes back to the beginning of the last century. An American engineer, H. Piper, patented a hybrid that used an electric motor to assist a gasoline internal-combustion engine. The unique selling proposition of this design was to propel the car up to the heady speed of 25 miles per hour. Unfortunately, straight gasoline engines were soon doing that on their own, leaving hybrids to join the other failed contenders—steam cars and battery-electric cars.

There are several types of hybrids. The simplest is a "stop-start" or "micro" hybrid. This is not really a proper hybrid, since when it is moving, it relies solely on the gasoline engine. It does, however, automatically shut down when the car is at a standstill, while an integrated starter-generator restarts the engine automatically when the accelerator pedal is pressed. Result: a 10 percent boost in efficiency and economy for no extra money. The next step up is the "mild" hybrid, such as Honda's Integrated Motor Assist (IMA), the configuration used in its early Insight hybrid and in Civic and Accord hybrid versions. The electric motor boosts the engine during acceleration. During braking, it acts as a generator, capturing kinetic energy that would otherwise be lost as heat on the brake pads. With the driver's foot on the brake pedal, the electric motor slows the car and charges the battery. The electric motor is always linked to the engine and so never propels the car itself. IMA is much cheaper than the complex Toyota system. In a Civic, it is supposed by independent experts to offer nearly 40 percent better fuel consumption than a gasoline model, but driving the early Insight models felt very awkward, with an oddly sluggish delivery of power and jerky motion when the accelerator was pressed, as if the engine was in too high a gear.

The early production versions of the Prius that Toyota showed in Britain, where the Japanese sedans fit in easily with their right-hand driving position, were equally unsatisfactory. Their futuristic shape turned heads and gathered lots of interested spectators whenever the author took his test vehicle to the local supermarket. But the handling was floppy, partly because the steering and suspension were tuned for Japan rather than Europe, where drivers expect the crisp, responsive steering you get with BMWs or Volkswagens.

Nevertheless, these turn-of-the-century, first-generation Priuses packed a technological marvel under the hood. Using a "power split" device, the output from the gasoline engine is divided and used both to drive the wheels directly and to turn the generator, which in turn drives the electric motor and also drives the wheels. So the distribution of power is continuously variable, allowing the engine to run efficiently at all times. When its full power is not needed to drive the wheels, the gasoline engine can spin the generator to recharge the batteries. The batteries also get recharged when the car is coasting or braking. In stop-and-go traffic and at low speeds, when the gasoline engine would be at its most inefficient, the engine simply shuts off, and the car glides along on the electric motor, powered by the battery. This leads the Prius to have better fuel economy in town (60 miles to the gallon) compared with on the highway, where it falls to about 50 miles per gallon.

People are flocking to buy hybrids. So how big is this wave? By the end of 2009, there will be seventeen manufacturers offering seventy-two models with hybrid engines, most of them variations of the complex Toyota full-hybrid system, often made under license from the Japanese pioneer. Of the total number of models, no less than 30 percent are Toyotas. Indeed, Toyota has about 77 percent of the hybrid market, measured by unit sales, followed by Honda, with 16 percent. Ford and GM have barely 2 percent between them.

Many automotive engineers grumble about the inherent inelegance of hybrids: why have a motor vehicle with two propulsion systems adding to the weight? Why indeed, except that such a system offers huge advantages in terms of extra fuel economy. Such dog-in-the-manger attitude is reminiscent of GM and Ford's failure to spot the rise of diesel in Europe, which led to their businesses there turning from cash cows to profit drains. Neither of Detroit's finest could see the writing on Europe's wall and to this day moan about their inability to extract from customers the extra $2,000 per car that a diesel engine costs. The fact that Peugeot, Renault, Volkswagen, Mercedes, BMW, and even Fiat could get their money back seems to have passed them by for a long time. Eventually, Ford caught up by doing a deal to share engine technology with Peugeot, while GM did at least profit on diesel engines from its otherwise disastrous dalliance with Fiat.

The thing about waves is that they form a pattern, with one following another. A wave of technology innovation is under way. And amid the turbulence, surprising newcomers are emerging. While the mainstream auto industry is moving ahead with the sorts of hybrids detailed earlier, a group of nerdy but influential enthusiasts in California is plugging into a radically different automotive future, using hybrid technology only as a starting point.

Plugging Into the Future

The license plate on Greg Hanssen's Toyota Prius turns heads: 100 MPG. Opening up the back, you see the secret that turns this Prius into something much more desirable: an ordinary electric plug. When Hanssen unveiled his plug-in Prius at an auto-industry convention back in the summer of 2006, he caused a sensation with his big idea. After he outlined how his hybrid could be recharged overnight from the grid back home in his garage, he sat down to a standing ovation. Auto engineers do not normally give ovations, but Hanssen delivered a message they all wanted to hear.

Like many great ideas, Hanssen's advance was simple. He hacked into a standard Prius's software, fitted it with a bigger, more robust battery, and ended up with a vehicle that could be used for most driving purely on battery power, with the gasoline engine simply there to be used as a booster or if the battery got run down. Otherwise, the battery could be recharged overnight, ready for the next day's journey.

So why does Toyota not offer the plug-in version itself? Why has it been left to hackers to mod their motors and usher in a bright new possibility for decarbonizing cars? The answer is because the Japanese firm worries that people will be turned off by cars that need to be plugged in. Toyota's sales pitch for the Prius is that you never have to worry about recharging, because the battery is recharged by the gasoline engine and by the kinetic energy captured during braking, when the electric motor reverses itself to become a generator. Toyota has concentrated on getting this straightforward message across and did not want to confuse it by having hybrids associated with pure electric cars, which flopped because of their lousy range.

But Hanssen's car has a superior battery, the lithium-ion technology found in laptops. He also hacked the software to prevent the gasoline engine from kicking in until the car is traveling at high speed. The result is that his modified Prius can do some 30 miles in all-electric mode, compared with less than 2 miles for a normal Prius. Even then, electric power is still blended in to improve fuel economy and provides up to three-quarters of the total power at 55 miles per hour.

Driving in Hanssen's car from San Diego to Los Angeles, you can see the payoff from this ingenious adaptation of a standard Prius. The onboard diagnostics screen on the dash verifies that the claim on the license plate is true: this car is getting 100 miles per gallon, nearly four times the fuel economy of the average new American car. The firm that converted Hanssen's car, Energy CS, has teamed up with Clean Tech, a systems-integration company, to offer plug-in retrofits for around $12,000 each. Toyota has been shamed and provoked into action and now promises plug-in hybrids within a couple of years—a promise that GM's vice chairman, Bob Lutz, echoed in 2006.

Who Revived the Electric Car?

But the significance of the Hanssen hackers goes beyond Toyota, beyond the Prius. It has revived interest in battery-electric cars just as Tesla Motors and other electric carmakers are poking their noses out of the ground. A film called *Who Killed the Electric Car?* released in 2006 by Sony Pictures, made a splash at the Sundance Film Festival—not least because GM and other automotive powers geared up a spin campaign against the film. The conspiracy-minded flick aimed to tell the "real" story behind the early demise of the GM EV1, an electric car introduced in the mid-1990s and withdrawn from the market by the car giant by 2004. Only a few hundred ever made it out into the real world.

Chris Paine, the film's director, and Dean Devlin, the executive producer (best known for *Godzilla* and *Independence Day*) both owned EV1s during the brief period a decade ago when the big car companies sold battery-electric cars in parts of America. The car giants did so not because they wanted to but because California regulators forced them to sell some zero-emissions cars. Grudging though it was, GM did manage to produce the superfast EV1—the most aerodynamic production car ever made. The EV1 proved hugely popular among California's green and gadget-loving set, the same crowd that is now rushing to buy the Toyota Prius. However, unlike today's hybrid cars, which are growing into a mass-market phenomenon, the EV1 and other electric cars bit the dust.

Car companies insist that the reason the original battery cars failed was lack of consumer interest, which has some legitimacy given the limited range of those cars, but that view is turned on its head here. The film investigates various possible culprits behind the "murder" of the electric car in turn—oil giants, carmakers, consumers, regulators, hydrogen energy (a rival technology), and so on—before pointing the finger at the true culprit. In one sequence, activists sneak into GM's secret testing grounds via helicopter and film the company crushing the beloved EV1s—in direct contradiction to the company's public vow to save the car.

Chelsea Sexton, a former marketer of EV1 cars and a star of the film, typifies the view of the plug-in crowd when she blames gullible regulators and cynical carmakers for abandoning electric cars for the distant dream of hydrogen. Inspired by the hacking of Priuses, various lobbying groups have sprung up, hoping to entice manufacturers to produce plug-ins and to push politicians to support them. Sexton, for example, now helps run Plug In America, a group that includes Jim Woolsey, a former head of the CIA.

Felix Kramer runs the California Cars Initiative (CalCars), a nonprofit advocacy group that promotes plug-ins. With help from Greg Hanssen's Energy CS, his outfit created the first plug-in Prius—though it used cheap lead-acid batteries, which are much heavier and shorter-lived than lithium-ion ones. During Earth Day celebrations in April 2006, Ron Gremban, CalCars' technology guru, led a group that converted a Prius into a plug-in in three days, while the public watched. In coordination with the Electric Auto Association, CalCars planned to release a free, open-source version of its conversion instructions.

Plug-In Partners, an entirely different group that counts many electric utilities and green groups as members, is drumming up "preorders" for fleets of plug-in vehicles to prove that a demand for them really exists. That is important not only because carmakers are notoriously risk-averse (given the huge sunk costs of existing capital stock). Battery enthusiasts whisper darkly that the car companies never wanted battery cars like the EV1 to succeed and so lied about a lack of consumer demand. Sexton and other former Detroit insiders point to long waiting lists they say were ignored by the big car companies, who chose instead to shut down their electric programs and to crush most of those electric cars.

They are united in thoughts: the advantages of adapting hybrids to operate in plug-in mode and the wickedness of the auto industry for ditching electric cars in favor of promoting the distant dream of fuel cells. The response of people at Toyota is to sigh about the cynical desires of the electric-power and coal industries trying to muscle into the transportation industry, which Big Oil and Big Auto have successfully carved up between them for over a century.

Hype or Hypercar?

If the Greg Hanssens and Chelsea Sextons pull off their revolution, though, that oligopoly will be busted and the great wall separating the giant industries of electricity and motorcars torn down at last. Once cars have the ability to connect to the grid to "download" power, there is no good reason that they cannot "upload" power, too, as micropower plants selling into the grid. This great convergence has been forecast by the energy guru Amory Lovins as the Hypercar, which he thinks will probably be powered by a stack of fuel cells.

Lovins has some intriguing ideas. In *Winning the Oil Endgame*, a book funded partly by the U.S. Department of Defense, he sketches out the mix of market-based policies that he thinks will transform the automobile and so propel the world to a good life after oil. Some of his supporters and even his staff, who tend to bleed green, are uncomfortable about working with the military, but Lovins is unapologetic. "I tend to work with just about everybody," he says. Why? "The world is shaped by lots of cultures, individuals, organizations. We can't expect to achieve much if we work only with a narrow slice." Besides, he notes, as a giant user of energy and polluter of the environment, the military is already a force in the energy world; helping it get greener surely helps the world's energy system move toward a more sustainable footing. The military also has deep pockets and is accustomed to long-term planning, something the political process tends not to do, so it takes deep thinkers like Lovins seriously.

That controversial stance taken by the Sage of Snowmass separates him from utopian environmentalists, but that is a nuance that sometimes escapes even the keenest of observers. In early 2007, the *New Yorker* ran a lengthy profile of Lovins by Elizabeth Kolbert, a brilliant and sensitive writer who often covers energy and environment issues, titled "Mr. Green: Environmentalism's Most Optimistic Guru." But is he really so green? And does he really see the world through rose-tinted glasses?

Lovins is undoubtedly concerned about the planet, but he rankles at the label. Conservation, per se, is not what interests him; it is the efficient use of resources. The former implies a moral change of behavior that results in fewer of the "energy services" (cold beer and hot showers, to use his phrase) made possible by energy, while the latter means squeezing out those services by using less energy in the first place. "I prefer elegant frugality, not a hair shirt," he says, distancing himself from a label "that has a lot of political baggage." As evidence of his ungreenness, he adds, "I don't mind taking an infinitely long shower at my home using solar energy and recycled water, because it doesn't waste anything." An inspection of that home in Old Snowmass, Colorado, which at 13,000 feet houses the world's highest banana farm, confirms that his environmental technologies do indeed allow him to have a very light ecological footprint. So well insulated is his large building, and so cleverly has he integrated renewable energy generators into it, that even in the coldest of winter months, it costs him only a pittance to heat the place; most of the time, in fact, his 1970s building is selling energy into the grid.

He also objects to being labeled an optimist, which in his mind implies the advocacy of unrealistic goals. Invoking an early mentor, the founder of Friends of the Earth, he explains, "David Brower taught me that optimism and pessimism are two sides of the same coin—both dangerous! Optimism is pejorative in its implication of irrational distortion, seeing the world through a rose-tinted lens that sees better outcomes than are feasible or realistic." Lovins certainly pushes unorthodox views, but his willingness to work for incremental change hand in hand with the military, oil companies such as Shell, and car giants such as GM does suggest that pragmatism, rather than optimism, is a fairer description.

And what of the notion that he is some sort of guru, as the New Yorker declared? For years, he used to rankle when such honorifics were uttered, be that guru or sage or visionary. As he approached sixty years of age, his reflexive modesty and self-deprecation at last began to give way to a justified acceptance of his status as a wise elder statesman. "Once or twice in a decade, I have a really big idea," he reflected. When pressed on why those really big ideas—be that the "soft" energy path, micropower, or light-weighting and the coming convergence of the car and energy industries—are usually not received with applause and recognition, he reflected before answering. If the world is really to change the energy system, he said, we have to start thinking fifty years ahead. And, he insisted, we must start by demolishing our preconceptions: "One can be 'unreasonable' without being *unreasoning*. I do feel constrained by the laws of physics, but not by what has been done before."

This rare mix of unorthodox thinking and pragmatism is evident in *Endgame*. Lovins first argues that America must double the efficiency of its use of oil through such advances as lighter vehicles. Then, he argues for a big increase in the use of advanced biofuels, made from homegrown crops, that can replace gasoline. Finally, he shows how the country can greatly increase efficiency in its use of natural gas in power generation, so freeing up a lot of that gas to make hydrogen for transport. That matters, for hydrogen fuel can be used to power cars that have clean fuel cells instead of dirty gasoline engines.

His strategy would end the century-long reign of the internal-combustion engine fueled by gasoline, ushering in the Hydrogen Age. And because hydrogen can be made by anybody, anywhere, from wind or nuclear power or natural gas, there will never be a supplier cartel like OPEC—nor suspicions from the antiwar crowd of "blood for hydrogen."

The most intriguing part of Lovins's argument is the notion that the automobile and energy industries, symbiotic twins and bitter enemies for over a century, will merge over time—to the benefit of both. For example, the increasing problem of grid breakdowns and power blackouts in North America and Europe could be tackled if today's creaking, centralized, and inefficient electricity grid were made smart enough to allow for micro-power plants, like fuel-cell cars. In the long run, merely building ever fatter pipes to supply ever more power from central power plants to distant consumers is not the solution to black-outs. Lovins explains why: "The more and bigger bulk power lines you build, the more and bigger blackouts are likely." A better answer is micropower—a great number of small power sources located near end users, rather than a small number of large sources located far away.

This sentiment is echoed by experts at America's Carnegie Mellon and Columbia universities, who have modeled the vulnerabilities (to trees or terrorists) of today's brittle power grid. Even the gurus at the Electric Power Research Institute (EPRI), which relies on funding from utilities that run big power plants, agree that moving to a distributed model, in conjunction with a smarter grid, will reduce blackouts. Look at Denmark, which gets around 20 percent of its power from scattered wind farms, for example. Skeptics argued that its reliance on micropower would cause more blackouts. It did not.

At first glance, this shift toward micropower may seem like a return to electricity's roots over a century ago. Thomas Edison's original vision was to place many small power plants close to consumers. However, a complete return to that model would be folly, for it would rob both the grid and micropower plants of the chance to sell power when the other is in distress. Rather, the grid will be transformed into a digital network capable of handling complex, multidirectional flows of power. Micropower and megapower will then work together.

The ABB Group foresees the emergence of microgrids made up of all sorts of distributed generators, including fuel cells (which combine hydrogen and oxygen to produce electricity cleanly), wind, and solar power. The University of California at Irvine is developing one now, as are some firms in Germany. "Virtual utilities" would then aggregate the micropower from various sources in real time—and sell it to the grid. Energy-storage devices will be increasingly important too. Electricity, almost uniquely among commodities, cannot be stored efficiently (except as water in hydroelectric dams). That means grid operators must match supply and demand at all times to prevent blackouts. But if energy could be widely stored on the grid in a distributed fashion and released cheaply and efficiently when needed, it would transform the reliability and security of the grid. The last few years have brought dramatic advances in this area. Several energy-storage technologies now look quite promising: advanced batteries, flywheels, and superconducting magnetic energy storage systems (SMES). But the most intriguing storage option involves hydrogen, which can be used as a medium to store energy from many different sources.

Most of the recent hoopla surrounding hydrogen has concentrated on its role in powering fuel-cell cars. However, its most dramatic impact may well come in power generation. That is because hydrogen could radically alter the economics of intermittent sources of green power. At the moment, much wind power is wasted because the wind blows when the grid does not need, or cannot safely take, all that power. If that wasted energy were instead stored as hydrogen (produced by using the electrical power to extract hydrogen from water), it could later be converted back to electricity in a fuel cell, to be sold when needed. Geoffrey Ballard of Canada's General Hydrogen and the former head of Ballard, a leading fuel-cell maker, sees hydrogen and electricity as so interchangeable on the power grid of the future that he calls them "hydricity."

Another benefit is that hydrogen could also be sold by wind farms to allow passing fuel-cell–powered electric cars to refill their tanks. In time, those automobiles might themselves be plugged into the grid. Tim Vail of GM calculates that the power-generation capacity trapped under the hoods of the new cars sold in America each year is greater than that of all the country's nuclear, coal, and gas power plants combined. Most cars are in use less than a tenth of the time. If even a few of them were plugged into the grid (in a parking garage, say), a "virtual utility" could tap their generating power, getting them to convert hydrogen into electricity and selling it to the grid for a tidy profit during peak hours, when the grid approaches overload.

This notion is not as far-fetched as it sounds. The Bay Area Rapid Transit (BART), San Francisco's public transportation authority, is looking closely at developing precisely such a "vehicle-to-grid" concept, which would allow commuters to park their micropower cars at BART garages, get automatically plugged into its microgrid, and sell power from the car into the grid as the system approaches its daily peak.

The embrace of Amory Lovins's vision would have two benefits. First, the collective power of all those clean micropower cars coming together to produce much-needed power just as all of California is turning on the air-conditioning could help douse spot electricity prices and thus act as the ultimate anti-Enron. Second, utility bills at the end of the month could turn out to be refund checks, with a note of thanks for helping prevent a blackout!

Of Fuel Cells and Fallacies

Larry Burns is GM's designated visionary. As the company's head of research and development and strategic planning, he is in charge of preparing the ailing automaker for a future in which oil gets more problematic and in which concerns over global warming will only grow. Before he rose to this job, his predecessors blew their early lead in electric cars and hybrid technology, giving away the short-term advantage to the Japanese. Some say Detroit is therefore out of the game, but Burns refuses to give up. He is throwing a Hail Mary pass that he is convinced will revive GM, and with it America's automotive prowess: he is betting his company's future on hydrogen fuel cells.

Although he first trained as a civil engineer, his technological curiosity has brought this engaging polymath a long way. By 1998, he had risen to the top of GM as its chief technology guru. Burns can charm any audience with his enthusiasm for technological advance. A viral illness in his midforties meant he woke up one morning almost completely deaf, but he overcame that with a cochlear implant, a sort of artificial ear that uses electronics to stimulate the auditory nerves. At one point in the early 2000s, Burns would carry a tiny microphone, which he held out to his interlocutor the better to catch his or her words. Now the only sign of his affliction is that he prefers not to hang out for conversations on the noisy stands of the Detroit Auto Show, preferring to retreat to a quiet room in the background.

Even as GM was heading into its darkest hours, he kept alive the flame of innovation so convincingly that Rick Wagoner and the board allowed him to spend $1 billion on developing fuel-cell vehicles. They and he knew that it would take at least twenty years before they became a significant part of car sales. But Burns was one of the first in the industry to come firmly to the view that one day, sooner or later, the internal-combustion engine will be replaced. He believes that only hydrogen-fuel-cell–electric vehicles will do the job without ruining the planet.

That is a surprising view from the heart of Detroit, especially from GM. But Burns is not alone: Bill Ford himself has pointed out that the century-plus reign of the internal-combustion engine is coming to an end. Burns's team has already made much progress in the search for the perfect car to break America's addiction to oil and to free Detroit from the grip of the internal-combustion engine. Burns's magic bullet is the fuel-cell car. It combines hydrogen with oxygen to create electricity that turns the wheels: the only thing that comes out of the tailpipe is drinkable water.

This has long seemed a will-o'-the-wisp, tantalizing but always out of reach, and possibly a wasteful distraction. Undaunted, Burns and the board of GM committed the company to bringing out a competitively priced fuel-cell sedan in 2010, one that could travel 300 miles on a tank of hydrogen fuel, twice the range of earlier prototypes by any carmaker. If a suitable network of hydrogen filling stations were in place, GM vowed to make a million of the revolutionary vehicles at prices to please ordinary Americans. To that end, GM announced in mid-2007 that it was moving fuel cells from its research labs to its development division. But will such a network shows materialize? Others have stumbled on similar territory.

In the early 1990s, Daimler-Benz (as it then was) started pouring 1 billion euros into fuel-cell vehicles that got their hydrogen from an onboard microchemical plant, which extracted the fuel from methanol. Both toxic and carbon-dioxide emissions were tiny compared with even the cleanest of gasoline cars. Fuel-cell versions of little Mercedes models impressed many as they did test runs from California to Washington, D.C. The company was going to have commercial versions on the market by 2004, but that deadline came and went, with only demonstration vehicles to show for all the R & D effort. Eight years after the start of the project, Daimler's charismatic boss, Jurgen Schrempp, had to admit defeat, and his fuel-cell guru, Ferdinand Panik, retired from the company. The problem was that no one was convinced methanol was the right pathway to hydrogen. Certainly the oil industry was not. As Bernie Bulkin, the chief scientist at BP at the time, said bluntly, "We don't like alternative fuels but the oil industry might change its fuel infrastructure once, for a zero-emissions fuel like hydrogen, maybe . . . but we'll be damned if we do it twice by switching first to a partly polluting, transitional fuel like methanol before getting to hydrogen."

That episode illustrates how tightly oil and autos are yoked together. One cannot move without the other. The auto companies might see huge benefits from fuel-cell cars one day in the future, not least because they would take their industry out of the whole environmental debate. But the oil industry is reluctant to abandon its birthright for an uncertain future, especially if it has to pick up the tab for the new infrastructure.

But while fuel-cell vehicles may be years away for the ordinary motorist, there are plenty of other alternative fuels and motors trying to treat the addiction to oil that afflicts America. As Bill Ford has pointed out, the undisputed mastery of gasoline and the internal-combustion engine is being challenged for the first time in a hundred years.

There is no single answer in the near term for drastically improving fuel consumption and cutting the emission of global-warming gases. Lightweight vehicles, advanced diesel and gasoline engines, hybrid gasoline-electric technology, powerful new lithium-ion batteries, and biofuels all have some part to play.

Away in the more distant future, say twenty years, the cost of fuel-cell–electric cars running on hydrogen could fall enough for the fuel-cell car to end more than a century of dominance by the internal-combustion engine. To confound skeptics who think this technology is but a dream, GM is working furiously to develop its potentially competitive model by the end of 2010. If Toyota is the hare sprinting ahead with the hybrid Prius, GM could one day be the tortoise winning in the long run with the fuel-cell car. Amazing as it may seem, the once-moribund automobile industry is beginning to regain the drive to innovate that lifted it to its earlier heights—what Amory Lovins applauds as "the spirit of Henry Ford and Ferdinand Porsche."

But what if the future belongs not to the incumbent giants of the industry but to nimble outsiders and radical thinkers? After all, breakthrough innovations tend to come not from within dinosauric industries but from irreverent and unpredictable entrepreneurs willing to break all the rules of the game. That is exactly what the instigators behind the Automotive X PRIZE are hoping.

The Race Is On!

Not so long ago, an unusual group of people gathered at the Googleplex in Mountain View, California, to plot the first great revolution of the twenty-first century. Larry Page, the young cofounder of the Internet search Goliath Google, convened the allday strategy session at his firm's headquarters, in Silicon Valley. This is no ordinary geek war game: Page is not gunning for his erstwhile nemesis Yahoo! or the perennial predator Microsoft. He is after much bigger game. He wants nothing less than to topple the car and oil industries, the twin pillars of twentiethcentury American capitalism.

The cabal includes some curious conspirators. One group, led by Vice President Al Gore, is outraged about the growing environmental damage done by SUVs and other gas-guzzlers that the American auto industry peddles. Another group, led by the former CIA director Jim Woolsey, is deeply worried about the national-security implications of America's rising dependence on oil from the Persian Gulf. Throw in a few leading venture capitalists, foundation moneymen, and Hollywood do-gooders into the mix, and you see why this group is so daring: they have the money, the political and cultural connections, and the entrepreneurial savvy to do an end run around the vested interests that have long dominated the car and oil businesses.

How, exactly? Enter Peter Diamandis, the brash but indomitable force behind the X PRIZE, an innovative charity that aims to bring about social and technological change by running contests with huge cash prizes to motivate breakthrough innovations. His inspiration was the Orteig Prize, which offered $25,000 for the first solo nonstop crossing of the Atlantic. Charles Lindbergh won that prize, but nineteen other competitors spent $400,000 trying—a flurry of investment that would otherwise probably not have taken place. The result was that the age of long-distance commercial aviation leapfrogged ahead by many years.

The first X PRIZE did the same for space flight. For decades, the scientists and engineers at NASA, McDonnell Douglas, Boeing, and other high churches of the space establishment insisted that space travel could be achieved only by governments. It was much too risky, capital intensive, and sophisticated for the private sector, they insisted.

Diamandis was convinced the status-quo powers were wrong. Thanks to his advanced aeronautics training from MIT, he knew there was no theoretical reason small spacecraft could not be designed and built. The obstacles were mostly commercial: nobody had done it, and the big boys had no interest in encouraging experimentation. So he hit upon the idea of a prize lucrative and glamorous enough to entice the world's best inventors and entrepreneurs. The first team of three people to launch into and return from space safely twice within two weeks would win $10 million. The response was breathtaking. Hundreds of applications poured in, and two dozen finalists from around the world gave it a serious try.

In the end, an American team funded partly by Paul Allen, a cofounder of Microsoft, won the prize—and thus was born a new age in space travel. By 2009, Virgin Galactic (a firm run by Sir Richard Branson, the swashbuckling British entrepreneur behind Virgin Atlantic Airways) plans to offer commercial, fare-paying passengers space flights from a new "spaceport" being built in New Mexico. Dozens of the first tickets, priced at $200,000, have already been snapped up. Branson is convinced that prices will fall within a few years to $50,000, about what people pay today for high-end trips to Everest or Antarctica.

Now, with the help of the Googleplex gang, Diamandis wants to repeat that trick with cars. He headhunted Mark Goodstein, a veteran of the legendary high-tech incubator Idealab. Together, they have come up with a new X PRIZE meant to spur the development of a car that can go a whopping 100 miles or more on a single gallon of gasoline. The organizers may even award a separate prize for the best "city car"—which might not accommodate five beefy Americans on a long road trip but would suit a daily commuter just fine—in order to encourage us to think differently about how we use cars. Since most Americans drive less than 25 miles a day, a two-seater commuter car just might take off as a family's second or third car if marketed well—and, of course, if the car itself is well made.

To make sure the winners don't turn out to be turkeys, the X PRIZE group envisions putting competitors through various tests of endurance, speed, and quickness. There is even talk of staging a nationwide race from California to Washington, D.C.—a sort of clean-energy *Cannonball Run*. The Hollywood types are buzzing about making a reality show out of the race. The winning car would not only have to make the journey swiftly and safely, it would also have to be desirable and built for mass manufacture. Millions of Americans would watch the show and vote online for their favorite car. The purse this time around: $25 million. "We are at a pivotal moment in time when promising technologies, growing consensus about climate change, and global politics make it ripe for a radical breakthrough in the cars we drive," argued Diamandis.

Is America ready to join the great global race to fuel the car of the future? As the next and final chapter of the book argues, the Great Awakening that is now spreading across the grass-roots of American politics is spilling over into its industries and financial markets, promising a clean-energy revolution even bigger than the telecommunications and Internet revolutions.

CHAPTER NINE

A Call to Arms

A grassroots movement sweeping across America
promises to overturn Washington's Oil Curse—
and level the playing field for clean energy
and the car of the future

Rent a Hummer the next time you visit the Motor City, and you'll get a glimpse of why things went so horribly wrong in Detroit. Pulling out of the Avis lot at Detroit's airport in a fire-engine-red model during a recent visit, one of the authors received an unexpected compliment: "Got the big boy today, sir? Way to go!" The story was much the same over the next few days, driving around town to smiles and approving waves of the sort that would be unimaginable even in other parts of America, let alone in other parts of the world.

The tragedy is that there is no greater symbol of Detroit's malaise than this SUV on steroids. GM unveiled its first Hummer at just about the time it decided to pull the plug on its EV1, the sleek, all-electric car that was sold only in California and parts of the American Southwest. In doing so, the firm gave away its lead in automobile electronics and thus its chance to pioneer the hybrid-electric technology that would later give Toyota a decisive advantage in the marketplace. Seduced by the short-term profits that SUVs offered during the boom years of the 1990s, GM and its American rivals neglected to invest in more fuel efficient, environmentally friendly technologies.

Worse than that, the Detroit firms actively lobbied for exemptions from fuel-economy standards and otherwise tried to manipulate government regulations to favor dirty, inefficient cars. The craze toward SUVs was made possible only because America's Corporate Average Fuel Economy (CAFE) law had a loophole that allowed vehicles that could be classified as light trucks to meet lower standards. Obviously, most people bought SUVs to use them not as commercial trucks but as passenger vehicles, and the loophole should have been closed—but Detroit's lobbyists in Washington fought tooth and nail against that.

This pattern of co-opting and perverting Washington's attempts at legislation and regulation is nothing new, of course: it is, in essence, a manifestation of the special Oil Curse that afflicts Washington. During the 1990s, Vice President Al Gore headed up the Partnership for a New Generation of Vehicles (PNGV), which included the Detroit carmakers but not their foreign rivals. The objective was to develop enabling technologies, such as hybrid-electric drives and better batteries, that would allow carmakers to produce "supercars" that got more than 80 miles per gallon. The carmakers were very reluctant to be pushed into making more efficient cars, despite the fact that taxpayers were funding this effort to the tune of $1.5 billion. They agreed to participate only on the condition that they would not be required to build any actual cars. The predictable result: the tax money was spent, but Detroit did not bring a single supercar into commercial production. Meanwhile, Toyota and Honda gave their brightest engineers the resources and top-level support they needed to develop the Prius and Insight hybrid cars.

The recent success (after ethanol's initial flop) of flex-fuel cars in Brazil also helps expose Detroit's pork-barrel relationship with Washington and the unintended consequences this can sometimes have. Eager to promote ethanol fuel as a renewable alternative to gasoline (and to help corn farmers in the Midwest), Congress demanded that Detroit carmakers make vehicles that could run on either gasoline or ethanol. The oil companies were dead set against ethanol, which they saw back then as a threat to gasoline. The American car companies didn't like this idea either, but they agreed to play ball if Congress would give them some relief from those dreaded CAFE standards as a reward for making such cars. They argued that flex-fuel cars would use less gasoline.

The result was an apparent success. Millions of such cars were sold, and Detroit gobbled up the CAFE credits. But it was a scam: many of the drivers of those cars did not even know they were flex-fuel cars, and there were few gasoline stations that sold E85 fuel (which is a blend of 85 percent ethanol and 15 percent gasoline) anyway. Those credits were therefore unjustified. The irony is that when Brazil got serious about promoting ethanol a few years ago, the big carmakers similarly complained and complained. Unlike America's Congress, though, the Brazilian government did not offer loopholes. When the carmakers had lost the final battle, they turned to the very same flex-fuel technology that America had developed and even improved it to the point that it now costs less than $50 per car. They solved the chicken-and-egg problem of engine and infrastructure facing any new fuel, and today most new cars in Brazil come equipped with flexible engines that can run happily on either fuel.

Brazil's fitful but ultimately successful ethanol experience, when combined with the innovative bus-rapid-transit systems pioneered by its city governments, shows that enlightened legislation and industry innovation can come together to support genuine alternatives to petroleum. Look at Washington, however, and you find the opposite: a cronyistic and craven political process and a shortsighted and superpowerful set of industries in Big Oil and Detroit. Just consider one more egregious giveaway that the car lobbies managed to wangle for the Hummer. As if the CAFE loophole were not offensive enough, Congress also passed legislation that gave the very heaviest of SUVs—read Hummers—extra tax breaks, supposedly to help small businessmen.

The power of the oil and car lobbies clearly keeps Washington addicted to oil. The good news is that Congress is not the only game in town when it comes to shaping the country's public policies—and Detroit's reckless embrace of gas-guzzlers is not shared by all. If you don't believe it, just drive that big red Hummer from Michigan to California, where America's energy and environmental future is unfolding, and see how much love your big boy gets.

Arnie's Green Brain

Arnold Schwarzenegger certainly makes an unlikely environmentalist. The tough-talking, cigar-chomping actor turned politician is better known for hogging Hummers than hugging trees. He has not one but half a dozen of the supersized SUVs—including some that are the original military versions, not the downsized H2 or cutesy H3 versions. Now, nobody can accuse Arnie of excessive political correctness. He continued sucking down cigars even in rabidly antismoking California, setting up a smoke tent on the grounds of the state capitol in Sacramento to indulge his passion. Nevertheless, even as the Governator kept puffing, he felt obliged to apologize for his fleet of gas-guzzlers. He even converted one of his Hummers to run on hydrogen made from clean renewable power.

A big part of the credit for the greening of the Governator must go to Terry Tamminen, a former businessman turned environmental advocate. Go back and read Schwarzenegger's original platform prepared as he ran for office in a special election, and you find a set of proposals that was praised even by leading environmental groups. Pitting economic growth against environmental protection is a false choice, declared the manifesto's opening line: Californians want and deserve both, and smart policies can accomplish both. What's surprising is that the author of that progressive platform was Tamminen, an energetic overachiever who looks more at home in a business suit than in Birkenstocks. That pathbreaking environmental platform helped nudge swing voters toward Schwarzenegger during his first election to office, but by the time it came time for his reelection campaign in 2006, polls confirmed that Arnie's green brain was the decisive factor sending him back to the state capital.

Visit Tamminen at the offices of Environment Now! an advocacy group in Santa Monica that he started some years ago, and the first impression one gets is that of a businesslike political operator. He is wearing a jacket with the crest of the state of California on the outside and a stitched inscription on the inside saying, "Terry, you are fantaaaaaastic! Arnie." Tamminen is a tall, sandy-haired man whose highly energetic personality seems to run on caffeine. He runs a business meeting in the outfit's conference room with the firm hand of a chief executive officer. But a discreet peek into his restroom confirms that this businessman turned Republican political adviser does indeed bleed green. Above the toilet is a bumper sticker that pleads,

"Give waterless urinals a chance. Build green. Everyone will profit."

The alternating images of greenbacks and greenery may seem schizophrenic, but it is precisely this dichotomy that makes Tamminen effective. His intuitive appreciation of the profit motive of business and his genuine passion for the environment helped him capture Arnie's imagination, and propelled the two of them to the vanguard of America's clean-energy revolution. When Schwarzenegger was running for office for the first time, he was casting about for good advice on environmental issues. During a Kennedy clan retreat back in Hyannisport, he asked Bobby Kennedy Jr., an ardent green, for advice. Kennedy insisted that he get in touch with Tamminen, his Californian friend who had founded Santa Monica Baykeeper, part of Kennedy's Water-Keeper alliance. Tamminen had also founded Energy Independence Now, an activist group that has lobbied for weaning California off oil by switching to hydrogen made from in-state renewable sources.

Tamminen readily concedes that he "wasn't born green" and explains that at the root of his transformation are his personal encounters with environmental degradation. As a youth, he was an avid scuba diver, enjoying the bountiful underwater life of the California coast. After a decade away in Europe and Australia, he returned to Los Angeles to relive his childhood Jacques Cousteau moments. To his shock, he found "a desolate wasteland of rock and mussels, of purple urchins and Styrofoam cups. No densely populated kelp beds, sculpted by the unseen hand of ocean currents, just a moonscape of barren rock covered in silt. What had happened to the lush kelp beds, the abalone, the sea bass, the lobster, the octopus?" That inspired him to found Santa Monica Baykeeper, and earned him his green credentials.

Unlike many in Washington, the men around Arnie are clearly not in the back pocket of the car or oil industries. But that's not all that motivated Schwarzenegger's unexpected green turn. He knew that Californians are proud of being the country's leader on clean energy. The state was the first to lead the charge against vehicular smog. California took a hard line against the car and fuel industries several decades ago, even before Congress passed the Clean Air Act in the wake of the first Earth Day nearly four decades ago. The state's regulators used a mix of incentives and technology-forcing mandates to push the two industries to introduce such breakthroughs as catalytic converters and cleaner fuels.

This did not come easily, however, as the oil and car industries fought regulators every step of the way. Terry Tamminen is stinging in his critiques of the dirty tactics, ranging from lawsuits to negotiating in bad faith to pushing faulty science, that he claims make the oil and car industries no better than Big Tobacco. In his stridently antifossil book, *Lives Per Gallon: The True Cost of Our Oil Addiction*, he makes this observation: "As a government official who has witnessed oil and auto industry duplicity and lobbying firsthand, I can testify that it is nearly impossible to perform that function sitting on the other side of the petroleum-powered smoke screen. Both the oil industry and the auto industry have acted again and again to deceive regulators about the hazards of their products and have used their wealth to hamstring attempts by state and federal legislators to make laws that address such threats."

But thanks to the courage and persistence of regulators like Tamminen and his predecessors and colleagues, the power of Big Oil and Detroit has been checked in California. As a result, the air is undoubtedly cleaner in Los Angeles than it was back in the 1970s when, in the words of one local politician, "it felt like a man was standing on your chest half the time." What's more, in solving the smog problem for L.A., the state's regulators have solved it for the rest of the country. Other American states typically wait for California to lead on an environmental initiative, and then follow along sometime later by adopting the same standards as their own. One can even argue that California's pioneering efforts on smog have helped the whole world leapfrog to cleaner technologies faster than otherwise possible. By doing nothing more difficult than copying Californian regulations, Chinese officials have guaranteed that the new cars manufactured outside Shanghai are significantly cleaner.

California has led in other ways too, especially in areas (like the power sector and buildings) where state regulations hold sway rather than federal laws (which typically govern automobile fuel-economy standards). Through a mix of carrots and sticks developed by the state Environmental Protection Agency, which Tamminen ran for a time, and the California Energy Commission, an efficiency-minded agency, the state has managed to develop greener buildings and more energy-efficient industries than much of the country. The astonishing result: the average Californian uses a whopping 40 percent less electricity than the average American. And that is not an artifact of a stagnant economy or shrinking population either, for the state's economy has grown handsomely over the last two decades.

Clearly, enlightened public policy works, even in America. The beauty of the California case history is that nobody can argue it is an irrelevant example from social-democratic Europe or communitarian Japan, where people somehow get by with those silly little Euro-cars or are packed into commuter trains by white-gloved official "pushers." On the contrary, Angelenos love to drive and invented much of car culture—as Tom Wolfe captured memorably in his first book, *The Kandy-*

Kolored Tangerine-Flake Streamline Baby. And even in northern California, which affects a greener sensibility than SoCal,
most people live not in Tokyo-style rabbit hutches but McMansions of the sort found all over the country. In other words,
when it comes to lifestyle, Californians are much like other
Americans—and yet they are able to enjoy the good things that
energy makes possible with energy efficiency that more resembles Sweden than St. Louis or San Antonio.

How has California accomplished what Washington clearly
has failed at? There are several explanations. First of all, as the
world's seventh-biggest economy in its own right, the state is
large enough to matter. If Louisiana or Connecticut (or Bolivia,
for that matter) had said that carmakers had to produce a certain number of zero-emissions cars or else be barred from the
state, Detroit would probably have laughed. The Golden State
is simply too lucrative a market for them to walk away. So even
as their lawyers sued to stop the state, they had their engineers
take the ultimatum seriously. Another reason California has been
able to lead is that it started tackling its smog problem so early
that Congress granted it a unique grandfather clause: California
and California alone has the right to adopt air-quality regulations that supersede the federal standards. All other states get to
choose the Washington standard or the California standard but
are forbidden from going a third way.

The main reason California has been able to make progress
on energy policy when Washington hasn't has to do with pork-
barrel politics. The car and oil lobbies have successfully stymied
every major attempt at reforming their industries that Washington has come up with, often invoking the national interest in
doing so, but their power in Sacramento has been checked. Of
course they are powerful in California too: Chevron is headquartered in San Ramon, and GM and Toyota have a joint

car-manufacturing facility in the state. But that has not proved enough leverage to hijack the political process.

Whenever these tired old industries claim that new policies will destroy jobs, Californians and their politicians are not automatically stuck in their tracks. On the contrary, Californians have for many years proved willing to accept some economic pain—in the form of higher gasoline prices and pricier new cars—in return for clearer air and cleaner energy. In other words, unlike Washington, California does not suffer from the Oil Curse.

Now comes the exciting part: California has decided to take on the twin challenges raised by automobile use of oil addiction and carbon, the federal government be damned. Arnie's Democratic predecessor, Gray Davis, had supported a groundbreaking law that would regulate the greenhouse gases emitted by automobiles sold in the state. The Bush administration joined Detroit and Big Oil in challenging that law, arguing that it was a backdoor way to set fuel-economy standards—which is a federal matter. When the governorship fell to a Republican ally of President Bush, many were convinced that the law would go by the wayside. Much to their shock, the new governor vowed to continue the fight to turn the law into a reality.

He went further, signing a law that demanded big carbon cuts from power plants supplying the state, even if they were located outside its borders, and committing California to reduce its net greenhouse-gas emissions by a stunning 80 percent below 1990 levels by 2050: "I say the debate is over, we know the science, we see the threat, and we know the time for action is now." He even gave his support for a carbon-trading system akin to the one the EU has implemented under the Kyoto Protocol and has suggested that his state might thumb its nose at the White House by joining that European system despite Bush's hostility to Kyoto. Tony Blair made a special visit to California in late 2006 to hug the Governator and praise his leadership on climate issues.

In inspiring his boss to charge ahead on global warming, Terry Tamminen was doing nothing less than taking on the global car and oil industries—and the Bush administration—head on. It is therefore not unreasonable to ask if perhaps Arnie's green brain has finally overreached. Could California's most recent attempts to tackle the nasty side effects of cars and oil be limited to the state, or fail altogether? On the contrary, they appear to have sparked a bottom-up movement across the United States.

Some might dismiss Arnie's green campaign as just another example of the Left Coast doing its own nutty thing. Schwarzenegger may be ostensibly Republican, but he is, after all, part of the Kennedy clan—and he does lead a state famous for Berkeley radicals, Birkenstocks, and bearded enviros.

The Great Awakening

On the crucial question of greenhouse-gas emissions from automobile tailpipes, New York and the New England states have indicated that they too will adopt those regulations if California's law is upheld in the courts. This is significant, because the governors of New York and Massachusetts (until 2006) who joined Arnie's campaign—George Pataki and Mitt Romney—were also Republicans. That suggests this is not merely a partisan attempt to do an end run around a Republican president. However, some have suggested the fact that the clamor for carbon regulation has been spearheaded by coastal states means that it is merely Bush-bashing: both coasts tend to be more liberal than the heartland, which is where much of the country's carbon-intensive coal deposits are found and where the politically powerful car and oil companies reside.

That may have been true a few years ago, but no more. The big surprise is that there is now an undeniable clamor for clean, carbon-free energy in the conservative heartland too. Thanks to regulations that favor renewable energy, oil-rich Texas is now one of the world's leaders in wind energy. Car-friendly Michigan is so keen to try to reinvent its decaying industrial base that the state government is pouring millions of dollars into Next Energy, a nonprofit consortium aiming to cultivate investments in grid micropower, fuel cells, alternative fuels, and other clean technologies. Even Chicago, the city of broad shoulders, has turned positively green under the environment and energy policies promoted by the Daley administration. The city has transformed abandoned and industrial lots into an unprecedented number of parks and green zones, and is encouraging developers to install micro wind turbines and solar panels atop the city's skyscrapers.

Politicians of all stripes representing America's Farm Belt are excited about the prospects for ethanol and renewable energy reviving their troubled agricultural economies. What is more, they are offering up some innovative and frankly surprising proposals. Tom Vilsack, the Democratic governor of Iowa until 2006, made his record as a clean-energy innovator a central theme of his first national speeches that tested the waters for a potential presidential bid in 2008. Though he ran a state built on corn, he dared to suggest that the current system of corn subsidies needs to be changed radically and that the unfair tariff keeping out cheaper Brazilian ethanol should be scrapped altogether.

Richard Lugar, a Republican senator from Indiana, another ethanol state, also dared to propose that the tariff on Brazilian imports be scrapped, for which he was branded a "traitor to the state" by political rivals. As head of the Senate Committee on Foreign Relations, he grew increasingly troubled by the geopolitical problems caused by oil, as well as its nasty environmental and health side effects. He held hearings on these hidden external costs involved in petroleum ("externalities," in the jargon of economists) and even dared to utter that four-letter word of Washington politics: "tax." Yes, he dared to call for a new petroleum tax, one that would kick in when the world oil price fell below $35 a barrel. Though economists quibbled about the unintended side effects of such a "floor price" policy, the fact that he had the courage to argue for raising taxes on gasoline surely deserves praise.

The clamor for meaningful action is not restricted to senators and governors. Over the last few years, cities and towns, municipal electricity cooperatives, nongovernmental groups, and other bottom-up organizations across the country have joined the chorus. Even leading evangelicals, including Billy Graham and Rick Warren (author of the best seller *The Purpose-Driven Life*), are now agitating for change. The presidents of thirty-nine evangelical colleges, heads of megachurches, and various aid groups and charities like the Salvation Army sent an open letter calling for President Bush and Congress to show leadership on the issue of climate change: "Many of us have required considerable convincing before becoming persuaded that climate change is a real problem and that it ought to matter to us as Christians. But now we have seen and heard enough." Revealingly, the religious leaders called for government to tackle global warming not with mandates issued from on high but through "cost-effective, market-based mechanisms."

The most surprising convert to the carbon and clean-energy cause is probably the Pentagon. For years, the military had been quietly funding cutting-edge research into fuel cells and hydrogen. That arose from a pragmatic concern about the limits of battery technology and the vulnerability of diesel supply lines to attack: top brass envision the "asset-light" army of the future, with climate-controlled bodysuits, ultralight digital computers, and superfast long-range vehicles that today's clunky batteries and hugely inefficient military engines cannot support.

But the twin threats to global stability posed by oil's carbon and concentration also began to worry the planners, as evidenced by two major studies they commissioned at the start of this new century. The Pentagon asked Peter Schwartz, a renowned futurist at the Global Business Network, to analyze the biggest potential threats facing America's supremacy in the twenty-first century—and he came up with global warming at the top of his list. The Pentagon also helped fund a major study by Amory Lovins, the efficiency expert at the Rocky Mountain Institute, on what it will take to break America's addiction to oil. Lovins showed in detail how a combination of greater fuel efficiency, lightweight materials for cars, biofuels, and related innovations could speed America through the "oil endgame."

As Thomas Friedman has observed in his *New York Times* columns, extraordinary "geo-green" coalitions have emerged in America: foreign-policy conservatives concerned with the geopolitical damage done by oil now join hands with environmentalist liberals worried about global warming. This matters for two reasons. First of all, the hotbeds of local activism on climate issues across the nation disproves the European stereotype of the ugly American, selfishly guzzling the world's resources while doing nothing about global warming. George W. Bush and Congress may dawdle on Kyoto, but add up all the bottom-up efforts and voluntary corporate initiatives, and you find that in fact, America has done rather more about climate than it appears, while Europe has frankly done less than it claims. It turns out a number of the EU countries prattling on self-righteously about Kyoto will not even meet their own self-imposed targets under that treaty.

The second big reason this grassroots revolution matters is that such shifting alliances and bottom-up activism are often harbingers of dramatic change in Washington's otherwise ossified political culture. From the civil rights movement that transformed politics in the 1960s to the Christian conservative movement that has defined much of American politics over the last two decades, change at the national level often starts with animated, agitated change at the local level. This is especially true when the emerging coalitions cut across partisan and geographic lines, as does the emerging movement to end America's oil addiction and tackle global warming.

The first Great Awakening of American politics to environmental issues that followed the first Earth Day back in 1970 showed this: the pathbreaking Clean Air Act and Clean Water Act were signed into office not by any bleeding-heart politician but by Richard Nixon. That proved the old adage among political scientists that Washington politics is a nonlinear system. For much of the time, vested interests fight each other to a bloody draw, so radical change is hard to pull off. This is especially true of big changes pushed from the top down. But, goes the argument, if the change is driven by genuine popular concern about a pressing issue, then even the most craven politician will be forced to turn away from pork-barrel politics for fear of losing touch with his or her constituency—and being booted out.

When the next Great Awakening hits Washington, change will come surprisingly quickly, as the old obstructionists turn opportunist and scramble to portray themselves as champions of the new political order. The local heroes of today's energy revolution are no longer content with local action; they are already knocking on Washington's door. Some of these unusual alliances have grown so emboldened that they are even advocating particular technologies that show special promise to address both national-security and environmental concerns quickly.

Austin Power

"Forget hydrogen. Forget hydrogen. Forget hydrogen!" That was the rallying cry of Jim Woolsey, a former director of the CIA, at an energy-technology event held at the National Press Club in Washington, D.C. He was referring to the idea, championed by Bush, that America might make itself less dependent on foreign oil by encouraging the development of hydrogen-powered cars.

Instead, the former spy chief has joined a curious coalition of environmental activists, national-security hawks, clean-energy experts, and politicians to unveil a national consumer campaign in favor of plug-in hybrid-electric vehicles. Another surprising supporter of plug-ins, Orrin Hatch, a senator from Utah and a conservative Republican not known for supporting green causes, also dropped in. Though the energy event took place at a crucial moment in the bitterly contested confirmation hearings of Samuel Alito to the Supreme Court, Hatch thought this issue important enough to run out of the hearings to give his two cents. He declared grandly to the plug-in crowd that this obscure technology could well be the "silver bullet" America needs to end its addiction to oil.

The event and the campaign it was designed to support were the brainchildren of Austin Energy, a power-generating utility owned by the city of Austin, Texas. Austin Energy's campaign has already won the endorsement of dozens of cities and towns, including Los Angeles, San Francisco, and Denver, as well as Austin itself, and also more than one hundred utility companies.

It planned to collect millions of signatures from individuals re-questing that big car firms start making plug-in hybrids. The idea behind collecting all of these "soft orders" is to prove there is vast market potential for ultraefficient cars—and to make sure that Detroit cannot claim, as it did after killing the first generation of electric cars, like GM's EV1, that nobody wanted to buy them.

Enthusiasts love plug-in technology because it would dramat-ically reduce oil use (which is why the national-security types are interested) and also curb greenhouse-gas emissions (which is why the environmentalists are interested). If it came to pass, it would radically restructure America's energy economics by shifting demand from the corner gasoline station to the power station. And who knows, it might even shift the global balance of another sort of power—the political variety.

Woolsey may well be wrong to bash hydrogen, but the for-mer CIA boss is certainly right in linking arms with a diverse, bottom-up coalition. Dealing with America's oil addiction is no longer a matter of concern only for policy wonks, just as dealing with climate change is no longer a matter just for greens. Washington can no longer afford to ignore this prob-lem because of this vibrant "coalition of the willing" made up of environmentalists, development experts, farmers, national-security experts, and religious leaders—or, as Woolsey puts it, "of tree huggers, do-gooders, sodbusters, cheap hawks, and evangelicals."

The Petro-Pragmatists and the Moon-Shooters

All the grassroots activism has clearly primed the pump, but America still needs Washington to act. Without a coherent na-tional energy and climate policy, there is a great risk that the blunderbuss of bottom-up efforts will ultimately fizzle out. Energy is not like other areas of public policy, like education or welfare, in which state-led experimentation by itself was enough to improve people's lives: electrons don't stop at the state border, and the carbon emitted from coal plants in Illinois can easily undo all the virtuous investments in clean energy made in California. That is why the next president and Congress will still have to come up with a twenty-first-century federal energy and environment strategy.

So what should such a plan look like? The energy-policy experts tend to fall into two camps. One group of gurus, call them the petro-pragmatists, argues that oil is bound to be with us for decades, and therefore we should abandon naïve talk of life after oil. They focus on steps that minimize the foreign-policy and domestic economic damage done by petroleum. The other group of wonks, call them the moon-shooters, argues that visionary leadership plus massive government investment can in fact wean the world off oil in a matter of years rather than decades, creating domestic jobs and leading to energy independence in the process. Each camp has some sensible things to say, but carry either agenda too far, and you end up getting nowhere fast.

Some outfits arguing a petro-pragmatist line are in fact fronts for dark elements of the fossil-fuel industry. Look closely at their policy proposals, which often emphasize "market forces" and a "need for further scientific research," and it becomes clear that they are really advocating a carbon-happy, do-nothing position. At the opposite end of the spectrum, some groups arguing for a new "Apollo Project" (a phrase meant to evoke JFK's bold vision of landing a man on the moon within a decade) or "Moonshot" to get America off oil within a decade through intense government action are equally wrongheaded. Look closely at the details of many plans for a "moonshot," and you find a confused coalition of labor unions, farmers, and environmentalists who cannot agree on much other than the need for massive government subsidies for their pet projects—be that making energy from turkey carcasses, used french-fry grease, or whatever other uneconomical or dirty approach boosts so-called energy independence.

The problem with both extremes of this argument arises from this oft-overlooked insight: markets and entrepreneurs, not governments, are usually the best agents to allocate capital and develop new technologies. That is, in essence, what is wrong with the extreme moon-shooter position. Sending a man in a tin can to the moon once, costs be damned, is a lot easier than transforming the entire energy and transportation system in a manner that is affordable, convenient, and sustainable. Government fiat alone can never accomplish that, no matter how many tens of billions of dollars or how much "vision" government puts into the job. On the other hand, the faux free-marketeers among the petro-pragmatic pack get it wrong too. Energy markets are not free anywhere in the world, least of all the petroleum market. OPEC tries to rig the world price through its cartel meetings, which is bad enough, but America's federal government doles out many billions in hidden subsidies to the oil and gas industries too.

That suggests there is in fact a legitimate government role in leveling the energy playing field so that clean-energy innovators can have a fighting chance in the marketplace against fossil fuels. One thoughtful report reflecting the intelligent petro-pragmatic view was put out by the Council on Foreign Relations (CFR). The CFR's energy task force, chaired by James Schlesinger and John Deutsch (both former cabinet officials and energy experts), argued that "America's dependence on imported energy increases its strategic vulnerability and constrains its ability to pursue foreign policy and national security objectives." The gurus offer various prescriptions for "managing the consequences of unavoidable dependence on oil and gas that is traded in world markets and to begin the transition to an economy that relies less on petroleum." The most powerful recommendations from the report concern ways to slow and ultimately reverse the growth in America's oil-guzzling: gasoline taxes, tougher and better fuel-economy laws for cars, and an innovative trading system for gasoline permits that would cap the total level of gasoline consumed in the economy.

Amory Lovins, head of the Rocky Mountain Institute, offers a moon-shooter counterpoint in his book *Winning the Oil Endgame*. He observes that "a modern car burns each day fuel derived from 100 times its weight in ancient plants; yet a mere 0.3% of that fuel moves the driver." The keys to the cars of the future, he argues, are to make them lighter, more efficient, and flexible-fueled, just as his team has done with its Hypercar: "Tripled-efficiency, ultra-light gasoline-hybrid sport utilities were designed in 2000, paying back in one year at European and Japanese fuel prices or two years at America's much cheaper pump prices. In 2007 the Automotive X PRIZE will start moving such designs to market. Just in America they will ultimately save 8 million barrels of oil a day—equivalent to finding a Saudi Arabia under Detroit."

Take the best ideas of the two rival camps, and it is possible to forge an energy strategy that is grounded in realism, but one that holds out hope for a brighter future. There is plenty of oil around, and it is likely to be cheap and accessible enough to be part of the energy mix for a long while—unless America acts decisively through public policy to ensure that the price at the pump reflects the true cost of oil. If it does that, then the more hopeful vision of disruptive innovations and life after oil could yet come about, though such breakthroughs will almost certainly arise not from the top-down diktats of bureaucrats but from the robust interplay of entrepreneurship and innovation.

Such a transition to life after oil will never happen unless Americans accept that artificially cheap gasoline simply keeps the country addicted to oil. This sea change in attitude is beginning to happen, as was made clear in a stunning interview given in late 2006 by Bob Lutz, vice chairman of GM and the ultimate Detroit power player. In offering a mea culpa for missing the hybrid-car trend, he made the most persuasive argument to take the Great Awakening seriously: "I think there's a bone-deep awareness in the American public now that $1.20 gasoline while the rest of the world is paying five, six dollars is not some God-given right because the maker decided to bestow cheap gasoline on the American public."

The Real Cost of Oil

Lutz is right. After all, Americans are forever complaining about prices at the pump. Now, it is true that Katrina sent pump prices shooting up across the nation, causing gas pains across the country in 2005 and early 2006. But polls showed that drivers were unhappy long before the storm bumped gasoline prices above $2 a gallon. Every summer, as America gets ready for what market analysts call the "driving season" leading up to Labor Day, consumers typically complain about prices.

But despite the perpetual moaning, gasoline is hardly over-priced. Adjust for inflation, and it turns out that Americans were paying as much or more for it a few decades ago. The $0.30-a-gallon price old-timers recall from the 1960s is, in to-day's prices, really about $2 a gallon; the $1.25-a-gallon price that more people will remember from 1980 translates in today's money to about $2.50 a gallon. In fact, gasoline is a bargain at today's prices. If you don't believe it, get out of your SUV and buy a quart of milk, a cup of java, or a bottle of mouthwash in the Kwikie Mart. Gallon for gallon, gasoline is still the cheapest liquid you can buy at most American filling stations.

Politicians in Washington have played up fears of high gasoline prices by launching investigations into price gouging by oil companies. The inquiries always come up empty. That is because Exxon and Chevron are easy to hate but in fact have little to do with setting oil prices: it is the OPEC oil cartel that has the power to rig the world oil market, not Western oil majors, and by far the biggest factor determining retail gasoline prices is the price of crude oil. Regardless, Americans still seem to feel that the price of gasoline should be only a buck a gallon.

The irony is that even the $2-plus price per gallon that is now upsetting Americans bears no relation to the true cost of gas. Securing, extracting, and burning petroleum brings with it all manner of complications, ranging from the geopolitical to the environmental, that cost the country dearly. The problem is that drivers do not pay for these things at the pump, so gasoline appears cheaper than it really is.

One sort of hidden cost arises from the militarization of America's energy policy. It is easy to pick on George W. Bush, the newly repentant oilman who sent troops to the oil fields of Iraq, but both parties deserve blame for perpetuating the country's addiction. It was FDR who first made the Axis of Oil deal with the Saudis after World War II, after all. And it was Jimmy Carter, the dovish conservationist, who first explicitly stated the military's mission to guarantee safe passage of oil from the Middle East "by any means necessary."

The Cato Institute, a libertarian think tank, calculates that America spent $30 billion to $60 billion a year safeguarding

Middle Eastern oil supplies during the 1990s, even though its imports from that region totaled only about $10 billion a year during that period. A more comprehensive study of oil's security subsidies added the costs of maintaining the taxpayer-funded Strategic Petroleum Reserve (a stockpile of crude oil kept as insurance against Middle East turmoil) and other oil-protection services (such as the Coast Guard clearing shipping lanes for, and providing navigational support to, oil tankers), and reckoned that the energy security subsidy for Big Oil is really $78 billion to $158 billion a year. Of course this form of subsidy is paid for by Americans as taxpayers, through the Pentagon budget, but because they do not pay for it at the pump, they enjoy the illusion that gasoline is cheap—and alternatives to oil face unfair competition in the marketplace.

There is another reason Americans do not pay an honest price for gasoline: tax breaks and pork-barrel giveaways for Big Oil. The scope, the scale, and the stealth of subsidies handed out to the oil industry are nothing short of shameful. There is a long tradition of giving it cheap access to oil on federal lands through tax relief. But this practice reached a grotesque excess with a decision made in the late 1990s. Claiming it was encouraging "energy independence" by boosting domestic oil sources, Congress waived royalties for firms drilling in part of the Gulf of Mexico. This giveaway cost taxpayers billions in lost royalties.

And yet the whole exercise was a sham. America, which consumes a quarter of the world's oil production every day but sits atop only 3 percent of its proven oil reserves, will never be energy independent as long as it uses oil. Even all the oil in the hugely controversial Alaskan National Wildlife Refuge (ANWR), indeed all the oil left under American soil, would not set the country free from the need to import oil. All this policy has done is line the pockets of oil companies that have recently been enjoying record profits anyway.

Arcane and obscure provisions in the tax code are another way that politicians help friends in the oil industry. Over the past few decades, the domestic oil industry has won the right to take overly generous depletion allowances, to expense its exploration and production costs, to earn credit for the costs involved in enhanced oil recovery, and to claim various other tax wheezes that are not available to other industries. The National Commission on Energy Policy, a sober, bipartisan panel of experts, argues that from 1968 to 2000, such giveaways have cost taxpayers the staggering sum of $134 billion or more (in inflation-adjusted dollars).

Worse yet, the handouts keep coming. Consider the Energy Act, which was passed into law in 2005. That monstrosity was one of the most pork-laden pieces of legislation to sail through Congress in years, prompting Senator John McCain to deride it as the "Leave no lobbyist behind" bill. Analysis done by Taxpayers for Common Sense, a watchdog group, suggests that the long-term cost of this bill could soar past $80 billion. Inevitably, the biggest share of subsidies went not to upstart clean technologies but to oil and nuclear power—mature, well-capitalized, and hugely profitable industries that deserve not one penny of government support.

The only thing more distressing is that fossil fuels also get a free ride in another way. The harm that these dirty fuels do when burned costs society and individuals a lot of money, but the driver does not pay for this at the pump. In the jargon of economists, the externalities of gasoline use are not rolled into its price. Instead, we pay through the suffering of asthmatic children, higher health-care costs, economic output lost to sickness, and damaged agricultural output, and, in coming years, through losses related to global warming too.

Some argue for ignoring such externalities, as they seem too fuzzy and immeasurable to matter. That lets oil off the hook too easily for two reasons. For one thing, ignoring such externalities unfairly disadvantages alternatives to oil that do not have such harmful side effects. Clean, homegrown alternatives like hydrogen fuel made from renewable energy sources or cellulosic ethanol made from agricultural waste would impose only tiny externalities on society—but they may never get a fair chance to compete, given that gasoline gets a free ride.

The second reason is that we do have some idea of the external costs. One study conducted in the 1990s by the University of California at Davis put the environmental and health externalities of burning oil at $54 billion to $232 billion a year. A separate study reported in the journal *Science* estimated the relevant health impact in Los Angeles alone to be almost $10 billion a year. Work funded by the Federal Highway Administration concluded that the total social cost of motoring could be in the hundreds of billions of dollars per year.

Given all this, it is only reasonable to ask what the real price should be at the pump for gasoline. Leading energy economists have argued that America's gas tax needs to increase by at least a dollar a gallon to account for externalities. Others insist the right figure is much higher. The honest answer is that there is no "right" price. Due to the uncertainty involved in estimating subsidies and externalities, nobody knows the answer with any precision. However, we do know two things with absolute certainty. First, while we may not know the precise size of oil's externalities, we can be sure it is not zero. Second, we know the direction gasoline prices must go in order to get closer to an honest price: up.

Level the Playing Field

The most sensible way to a clean energy future is to unrig the game. Americans should start by naming and shaming the companies that benefit from subsidies, expose the amount of giveaways, and shine a fiery spotlight on the politicians who steal from ordinary taxpayers in order to keep the lobbyists happy. And when courageous politicians—Republicans like Richard Lugar and John McCain or Democrats like Al Gore and Christopher Dodd—dare to speak about the true cost of oil or the need for subsidy reform and carbon taxes, citizens should stand up and applaud these brave souls. And vote for them. And encourage their friends and communities to rally behind such folk.

The reason politicians are so reluctant to do this, of course, is because of what Pete Peterson calls the turkey rule. "I grew up in Nebraska," says the billionaire chairman of Blackstone Group, a leading private-equity firm, "and back there the first turkey to stick its head up from the pack got its neck chopped off." Politicians today are the same, argues the feisty octogenarian, in their refusal to deal with America's "energy gluttony." Perhaps emboldened by his private fortune or his advanced years, he proclaims that alternative energy will take many years to mature and therefore the only answer in the meanwhile is to consume less. To do that, he proposes a "very large tax on oil" that would lift American pump prices even higher than European ones today, and much tougher fuel-economy standards for new cars. He fears that if anybody running for office made such a speech today, "it would probably be political suicide."

"Tax" has long been a dirty word in American politics, and energy subsidies remain deeply entrenched in the pork-barrel Washington process. Jay Inslee, a congressman from Washington State, agrees with Peterson that change will not come

easily: "D.C. is a castle trying to resist good ideas and technology coming from all over the U.S.A. The federal government
has a defeatist attitude." But as a paid-up moon-shooter (he is
a supporter of a clean-energy advocacy group called the Apollo
Alliance), he firmly believes the system can change. In an interesting convergence between the two camps of thought, he
argues the key reform will be "real prices or caps on carbon
emissions justified by externalities."

Change can come in the American political process, especially when new coalitions and shifting alliances press for such
change. And when that happens, as may be the case with today's
geo-green alliances, the change could come mighty fast.

A decade ago, most Americans were completely unaware
that agricultural subsidies in America and Europe were as huge
or as destructive (both to Washington's finances and to Africa's struggling farmers) as they are. Today, thanks to a noisy
campaign waged by development activists and market-minded
economists, you can go on the Internet (www.ewg.org), put in a
zip code, and see exactly how much subsidy agribusiness giants
and movie stars with trophy ranches get in your neighborhood.
The vanguard of this next political revolution has also set up a
Web site for naming and shaming energy subsidy abuses (www.
greenscissors.org). As Americans come to understand how huge
the cash pile given to Big Oil is, they are sure to clamor for
reform.

Equally important, though undoubtedly more politically difficult, is the need to tax the "external" costs of oil. There is
more than one way to do this, but an economy-wide "carbon
tax" would be the simplest and most efficient approach. Such
a tax, imposed on fuels that produce greenhouse gases (such
as carbon dioxide) when burned, would be seen by consumers
in effect as a gasoline tax. Because this approach would be

economy-wide, it would not discriminate against only the power sector, as current legislation at the state level does. Nor would it fumble about in bureaucratic fashion with mandates, as the CAFE fuel-economy standards on cars do.

A carbon tax, slowly phased in, would send a clear signal to markets that the externalities of burning fossil fuels matter. A tax on gasoline might even win Detroit's support if the CAFE law is scrapped or reformed. And if the revenue collected from such a tax is returned to households quickly as income-tax re-fund checks, the next president and Congress could overcome suspicions that this is just a ruse to feed Big Government. In fact, given the geopolitical benefits, they could even market it as the Patriot Tax Refund.

Without picking winners, such a policy would also spur in-novation and investment in clean-energy technologies such as fuel cells. The good news is that a promising suite of technolo-gies—ranging from flex-fuel ethanol engines to plug-in hybrids to hydrogen fuel cells—finally offers a way to move beyond oil and the internal-combustion engine. As these technologies take off, they could even mean the death of OPEC. Because biofuels and electricity and hydrogen can be produced anywhere, from a wide range of energy sources, this brave new energy world could never be held hostage by terrorists.

Of course, there is no magic level for such a tax. Just as gov-ernment cannot possibly know what is the right technology to replace oil, neither can it possibly know what is the right price for gasoline. The best tax policy would send a bipartisan, long-term signal to the markets that the externalities of fossil-fuel use matter. For example, the next president could announce a bipartisan plan to raise the price of gasoline by, say, a nickel or a dime per gallon once a month, every month, for twenty years. The simple, clear price signal would resonate across the economy—rewarding clean energy and forcing dirty energy to pay its way for the first time.

A dime may not seem like much, but the fact that the gradual start is part of a long-term, predictable trend will influence individuals and firms. Families might decide that they will spring for a slightly smaller SUV the next time they buy a car, given the predictable direction of gas prices, and Detroit may decide to accelerate rather than kill its investments in clean technologies. Though businesses, especially the oil business, will attack this as a Big Government approach, it is exactly the opposite: unlike blunt regulations such as CAFE or EPA rules that mandate specific technologies, this elegant and simple approach leaves the marketplace to pick the technology winners.

The obvious objection to this fine theoretical notion is that Americans will never, ever support a rise in gasoline taxes, right? Think again. A *New York Times* poll published in early 2007 showed that when you ask if they support a rise in gasoline taxes, most Americans do say no. But ask the question again, and explain that the gas tax would help cut reliance on imported oil, and something truly astonishing happens: a majority of respondents say yes. That's right—the majority of respondents told pollsters they would actually support an increase in the gasoline tax. And that poll did not even tempt them by offering the smarter revenue-neutral tax that would put money right back into their pockets.

In short, there is an opening for leaders with vision—but they desperately need support. They certainly aren't getting it from most mainstream environmental groups in America, who have demonstrated unpardonable political cowardice when it comes to advocating "eco-taxation." Two decades of success in Europe shows that economies can flourish while improving the environment by shifting taxes away from "good" things government wants to encourage, like labor or social security, and onto "bad" things like carbon pollution or chemical waste. Or if a leader wanted to sell a tax more easily in America, he or she could adopt that idea of a Patriot Tax Refund, which would make any gasoline tax "revenue neutral" by handing back refund checks to every household—though obviously households that drove three Hummers would pay more for gasoline with the new tax than ones that drove Priuses or bicycles.

That proposal should be at the top of the short list of smart energy policies that the next American president and Congress should put in place. Here are five first principles that such a twenty-first-century energy policy should follow:

1. *Americans need to pay honest prices for fossil fuels:* The cost of gasoline must reflect the true cost to society imposed by its environmental, geopolitical, and economic harm. The best way to accomplish this is through revenue-neutral taxation and the elimination of subsidies, which would level the energy playing field so that clean alternatives finally have a fighting chance.

2. *The business of business is business:* Don't expect corporations to act out of goodwill, charity, or "corporate social responsibility" to tackle oil addiction. There is nothing immoral or surprising about oil companies selling oil or car companies selling SUVs, and voluntary schemes and claims of being "beyond petroleum" should be discounted. If Americans want companies to move beyond oil, they must change the social contract through government action.

3. *Leave it to the market to pick the winners:* The temptation is strong, especially among moon-shooters, to

look to the government to back promising technologies. However, history shows this is a formula for disaster. No group of officials, no matter how benevolent or well funded, can match the dynamism of markets and entrepreneurs in coming up with innovative technologies and business models that best meet consumers' needs.

4. *Government must act:* While bureaucrats should not push favored technologies, the conventional laissez-faire argument for government to do nothing falls short. There is a clear case for government intervention in energy and environmental policy due to the costly externalities involved in burning fossil fuels. In addition to externalities pricing, there is a strong case for specific regulations (such as a market-minded substitute for the CAFE fuel-economy standards) and especially for investment in the much-neglected areas of technical education and basic energy research.

5. *Individual action is the essential catalyst for change:* The key to driving change in America's political system is grassroots rebellion. As individuals and communities come together as part of this Great Awakening to demand better from the country's leaders, political leaders of vision will at last have the chance to step forward and answer their call.

If the next president and Congress have the courage to craft a new energy policy based on this market-minded manifesto, America may yet prove the petro-pragmatists wrong by leaving oil behind long before the world runs out of it.

A Call to Arms

The authors have grilled American environmental leaders for a decade about why they remain silent on the need for taxes on dirty energy, which most agree privately are a good idea. The head of one leading green group known for its support of other market-based policies hung his head and said, "Yes, a carbon tax would be the best way forward, but the moment we say the word 'tax,' we'll lose 100,000 members." Senior members of another giant environmental group laughed at one of the authors over dinner one evening about how politically naïve it was to think they might support higher energy taxes—but one of them later sent a private e-mail saying he supported the idea and hopes his group can do better in future.

Philip Sharp, a former congressman who now runs Resources for the Future, a respected environmental think tank, believes firmly in the case for externalities policies but knows from first-hand experience how hard they are to pass through Congress. He voted for modest gasoline-tax increases in the wake of the 1970s oil shocks but points out that "this was not so difficult back then because the money went to build roads" (yet another perverse subsidy for cars and oil, since the lack of proper road pricing in America means that drivers do not pay the true cost of motoring). He recalls that when Bill Clinton tried to pass a real energy tax early in his first term, even the Sierra Club, which initially made positive noises about the idea, ultimately refused to support him.

The green paralysis is a disgrace. After all, how on earth can politicians, who must actually face the slings and arrows of angry voters every few years, be expected to have the courage to advocate higher energy taxes when environmental leaders do not speak out first to educate voters and to give those elected officials political cover? Green leaders in Europe, for all their flaws and hyperbolic tendencies, at least have done a decent job of helping European consumers understand that fossil fuels—especially oil—have vast hidden costs and that paying an artificially cheap price for them at the pump only perpetuates petro-addiction.

Because the green movement in America has not yet shown leadership on this matter, it is all the more important for individual voters to support politicians at the local, state, and national level who advocate creating a genuinely level playing field in energy. But even here, there are signs of change. Consider the extraordinary mea culpa offered by one of the most respected elders of America's environmental movement. Gus Speth was one of the founders of the Natural Resources Defense Council and previously led several high churches of the green order, including the World Resources Institute and the United Nations Development Programme. Reflecting on a career in environmental policy and activism from his current perch atop Yale's School of Forestry and Environmental Studies, he had this to say: "Thirty years ago, the economists at Resources for the Future were pushing the idea of pollution taxes. We lawyers at NRDC thought they were nuts, and feared that they'd derail command-and-control measures like the Clean Air Act, so we opposed them. Looking back, I'd have to say this was the single biggest failure in environmental management—not getting the prices right."

Astonishing as it must seem, Detroit is now publicly advocating carbon constraints. During congressional hearings on global warming held in early 2007, both the Big Three American carmakers and the United Auto Workers supported efforts to cap greenhouse-gas emissions from cars. John DeCiccio of Environmental Defense declared that this was "a major breakthrough." On the heels of that news came another stunner: the Supreme

Court ruled against the Bush administration on a crucial case involving regulation of tailpipe emissions of greenhouse gases. This ruling paves the way for California and others to tackle global warming from cars in earnest, and may yet come to be seen as the Roe v. Wade moment of climate change.

Clearly, something really big is happening in energy politics. Talk at last is turning to honest pricing for gasoline, and entirely new coalitions are forming. Margaret Mead said, "Never doubt that a small group of thoughtful, committed citizens can change the world. Indeed, it is the only thing that ever has." Entrenched dinosaurs like Big Oil and Detroit have no problem lobbying their way around conventional political obstacles, but they tremble when citizens organize themselves around a powerful vision and mobilize. As the remarkable life of Terry Tamminen, the scuba diver turned market-minded environmentalist, demonstrates, the actions of grassroots groups and individuals like the readers of this book can make all the difference.

If Americans do come together and demand much better from the country's leaders, then the world has every chance of moving beyond oil. The future may yet see Sheik Yamani, the former Saudi oil minister who shot to global prominence during the oil shocks of the 1970s, proved right when he said, "The Stone Age did not end for lack of stone, and the Oil Age could well end long before the world runs out of oil."

The race is on to fuel the car of the future. From Silicon Valley to Shanghai, policy and technology innovators are zooming ahead of Big Oil and Detroit in reinventing the juice and the jalopy. While the dinosaurs dawdle, giants from other industries—ranging from GE to DuPont to Wal-Mart to Virgin—are swooping in, hoping to seize the clean-energy future. As they do so, they help answer Mahatma Gandhi's question about how many planets it will take if Asia follows the rich world's path of industrialization: we can live within one planet's worth of resources as long as we nurture and tap the limitless fount of human ingenuity.

If we see that oil is the real problem at the heart of our energy and environmental dilemma, then cars can be part of the solution. The altogether new coalitions made possible by envisioning the clean car of the future as a driver of change makes it possible at last to see a path to life after oil.

Bibliography

Baer, Robert. *Sleeping with the Devil: How Washington Sold Our Soul for Saudi Crude*. New York, Crown, 2003.

Bak, Richard. *Henry and Edsel: The Creation of the Ford Empire*. Hoboken, NJ, John Wiley & Sons, Inc., 2003.

Brady, Chris, and Andrew Lorenz. *The End of the Road: BMW and Rover—A Brand Too Far*. London, Pearson Education Limited, 2001.

Bronson, Rachel. *Thicker Than Oil: America's Uneasy Partnership with Saudi Arabia*. New York, Oxford University Press, 2006.

Cannon, James. *Harnessing Hydrogen: The Key to Sustainable Transportation*. New York, Inform Inc., 1995.

Curcio, Vincent. *Chrysler: The Life and Times of an Automotive Genius*. New York, Oxford University Press, 2001.

Deffeyes, Kenneth. *Beyond Oil: The View from Hubbert's Peak*. New York, Hill and Wang, 2006.

Easterbrook, Gregg. *A Moment on the Earth: The Coming Age of Environmental Optimism*. New York, Penguin, 1996.

Esty, Daniel, and Robert Winston. *Green to Gold: How Smart Companies Use Environmental Strategy to Innovate, Create Value, and Build Competitive Advantage*. New Haven, CT, Yale University Press, 2006.

Faith, N. *Crash: The Limits of Car Safety*. London, Boxtree, 1997.

Feast, Richard. *Kidnap of the Flying Lady: How Germany Captured Both Rolls-Royce and Bentley.* St. Paul, MN, Motorbooks International, 2003.

Gallagher, Kelly Sims. *China Shifts Gears: Automakers, Oil, Pollution, and Development.* Cambridge, MA, MIT Press, 2006.

Ghosn, Carlos, and Philippe Ries. *Shift: Inside Nissan's Historic Revival.* New York, Doubleday, 2005 (translated by John Cullen, originally *Citoyen du Monde.* Bernard Grasset, France, 2003).

Goodstein, David. *Out of Gas: The End of the Age of Oil.* New York, W.W. Norton and Company, 2005.

Gregor, Neil. *Daimler-Benz in the Third Reich.* London, Yale University Press, 1998.

Hiraoka, Leslie. *Global Alliances in the Motor Vehicle Industry.* Westport, CT, Quorum Books, 2001.

Holweg, Matthias, and Pil, Frits. *The Second Century: Reconnecting Customer and Value Chain Through Build-to-Order.* Cambridge, MA, MIT Press, 2004.

Howard, George. *Stan Ovshinsky and the Hydrogen Economy: Creating a Better World.* Notre Dame, IN, Academic Publications, 2006.

Ingrassia, Paul, and Joseph White. *Comeback: The Fall and Rise of the American Automobile Industry.* New York, Simon and Schuster, 1994.

Kewley, S. J. *Toyota's French Connection: Trends in Japanese-European Automotive Relations.* London, Royal Institute of International Affairs, 2002.

Kiley, David. *Driven: Inside BMW, the Most Admired Car Company in the World.* Hoboken, NJ, John Wiley & Sons, Inc., 2004.

Kleveman, Lutz. *The New Great Game: Blood and Oil in Central Asia.* New York, Grove Press, 2004.

Kynge, James. *China Shakes the World: A Titan's Rise and Troubled Future—and the Challenge for America*. Boston, Houghton Mifflin, 2006.

Leggett, Jeremy. *The Empty Tank: Oil, Gas, Hot Air, and the Coming Global Financial Catastrophe*. New York, Random House, 2005.

Levin, Doron. *Behind the Wheel at Chrysler: The Iacocca Legacy*. New York, Harcourt Brace and Company, 1995.

Liker, Jeffrey. *The Toyota Way: 14 Management Principles from the World's Greatest Manufacturer*. New York, Mc-Graw-Hill, 2004.

Lutz, Robert. *Guts: 8 Laws of Business from One of the Most Innovative Business Leaders of Our Time*. Hoboken, NJ, John Wiley & Sons, Inc., 2003.

MacKenzie, James; Dower, Roger and Chen, Donald. *The Going Rate: What It Really Costs to Drive*. Washington, World Resources Institute, 1992.

Maddison, Pearce, et al. *Blueprint 5: The True Costs of Road Transport*. London, Earthscan Publications Limited, 1996.

Madsen, Axel. *The Deal Maker: How William C. Durant Made General Motors*. New York, John Wiley & Sons, Inc., 1999.

Magee, David. *Turn Around: How Carlos Ghosn Rescued Nissan*. New York, HarperCollins, 2003.

Magee, David. *Ford Tough: Bill Ford and the Battle to Rebuild America's Automaker*. Hoboken, NJ, John Wiley & Sons, Inc., 2005.

Mast, Tom. *Over a Barrel: A Simple Guide to the Oil Shortage*. Austin, TX, Hayden Publishers, 2005.

Maxton, Graeme, and John Wormald. *Time for a Model Change: Re-engineering the Global Automotive Industry*. Cambridge, UK, The Cambridge University Press, 2004.

Maynard, Micheline. *The End of Detroit: How the Big Three*

Lost Their Grip on the American Car Market. New York, Doubleday, 2003.

Nadis, Steve, and James MacKenzie. *Car Trouble: How New Technology, Clean Fuels, and Creative Thinking Can Revive the Auto Industry and Save Our Cities from Smog and Gridlock.* Boston, Beacon Press, 1993.

Nieuwenhuis, Peter Cope and Janet Armstrong. *The Green Car Guide.* London, Green Print, 1992.

Nieuwenhuis and Peter Wells (eds.). *Motor Vehicles in the Environment: Principles and Practice.* Chichester, John Wiley & Sons, Inc., 1994.

Odell, Peter. *Why Carbon Fuels Will Dominate the 21st Century's Global Energy Economy.* Essex, UK, Multi-Science Publishing Co., 2004.

Olah, George A., Goeppert, Alain, and Surya Prakash, G. K. *Beyond Oil and Gas: The Methanol Economy.* Weinheim, Wiley-VCH, 2006.

Organisation for Economic Co-operation and Development (OECD) and The European Conference of Ministers of Transport (ECMT). *Urban Travel and Sustainable Development.* Paris, ECMT, OECD, 1995.

O'Toole, Jack. *Forming the Future: Lessons from the Saturn Corporation.* Cambridge, MA, Blackwell Publishers, Ltd., 1996.

Pucher, John, and Christian Lefvre. *The Urban Transport Crisis in Europe and North America.* London, Macmillan Press Ltd., 1996.

Rifkin, Jeremy. *The Hydrogen Economy: The Creation of the Worldwide Energy Web and the Redistribution of Power on Earth.* New York, Tarcher/Putnam, 2002.

Roberts, Paul. *The End of Oil: On the Edge of a Perilous New World.* Boston, Mariner Books, 2005.

Schiffer, Michael. *Taking Charge: The Electric Automobile in America*. Washington, Smithsonian Institution Press, 1994.

Setright, L.J.K. *Drive On! A Social History of the Motor Car*. London, Granta Books, 2002.

Simmons, Matthew. *Twilight in the Desert: The Coming Saudi Oil Shock and the World Economy*. New York, John Wiley & Sons, Inc., 2005.

Stavins, Robert. *Economics of the Environment: 4th ed.* New York, W.W. Norton, 2000.

Tamminen, Terry. *Lives Per Gallon: The True Cost of Our Oil Addiction*. Washington, Island Press, 2006.

Toyota Motor Corporation, *Toyota in the World: 2006*.

Vaitheeswaran, Vijay. *Power to the People: How the Coming Energy Revolution Will Transform an Industry, Change Our Lives, and Maybe Even Save the Planet*. New York, Farrar, Straus and Giroux, 2003.

Vlasic, Bill, and Bradley Stertz. *Taken for a Ride: How Daimler-Benz Drove Off with Chrysler*. New York, HarperCollins, 2000.

Waller, David. *Wheels on Fire: The Amazing Inside Story of the DaimlerChrysler Merger*. London, Hodder and Stoughton, 2001.

Wells, Peter, and Michael Rawlinson. *The New European Automobile Industry*. New York, St. Martin's Press, Inc., 1994.

Williams, Richard. *Enzo Ferrari: A Life*. London, Yellow Jersey Press, 2001.

Wolfe, Tom. *The Kandy-Kolored Tangerine-Flake Streamline Baby*. New York, Bantam, 1999.

Wollen and Kerr (eds.). *Autopia: Cars and Culture*. London, Reaktion Books, 2002.

Womack, Jones and Roos. *The Machine That Changed the World*. New York, Rawson Associates, 1990.

Yergin, Daniel. *The Prize: The Epic Quest for Oil, Power, and Money*. New York, Free Press, 1993.

Index

Acknowledgments

We are both very grateful to John Micklethwait, *The Economist*'s editor in chief, for giving us time to work on this book as well as permission to use articles that we have written for the magazine. We thank Bill Emmott, his predecessor, for his many years of support and early encouragement of this project. Grudgingly, we also thank Daniel Dombey for correcting our steering at a crucial moment. The authors also thank Michael Coulman of *The Economist*'s research department for his help with this book.

About the Authors

Iain Carson has been the industry editor of *The Economist* since 1994, covering the airline, transportation, and manufacturing industries. He has also worked as a reporter and anchor for BBC Television and Channel 4.

Vijay V. Vaitheeswaran is an MIT-trained engineer and correspondent for *The Economist* with a decade of experience covering environmental and energy issues. He teaches at NYU's Stern School of Business and is a term member at the Council on Foreign Relations. He is also the author of *Power to the People*.

About TWELVE

TWELVE was established in August 2005 with the objective of publishing no more than one book per month. We strive to publish the singular book, by authors who have a unique perspective and compelling authority. Works that explain our culture; that illuminate, inspire, provoke, and entertain. We seek to establish communities of conversation surrounding our books. Talented authors deserve attention not only from publishers but from readers as well. To sell the book is only the beginning of our mission. To build avid audiences of readers who are enriched by these works—that is our ultimate purpose.

For more information about forthcoming TWELVE books, you can visit us at www.twelvebooks.com.